Pilot Competency and Capability

Pilot Competency and Capability presents strategies for the air carrier pilot-in-command operating complex engineered systems within a complex natural environment. It bridges the gap between academic books and practical application by providing real-world examples of how various safety and operational theories work in practice.

The book advises on how to develop concepts, strategies, and ways of thinking that integrate with existing structures and FAA regulations, while understanding how engineered systems and codified structures interface with complex natural environments. It considers how the prescribed safety margins function to manage emergent behaviors of both the natural environment and the engineered systems.

The book is intended for airline pilots, training captains, simulator instructors, and aviation students taking courses in aviation safety, risk management, and flight safety to improve in-flight decision-making, risk analysis, and strategic planning.

Pilot Competency and Capability

Responsibilities, Strategy and Command

Steven D. Green, FRAeS

CRC Press
Taylor & Francis Group
Boca Raton London New York

CRC Press is an imprint of the
Taylor & Francis Group, an **informa** business

Designed cover image: iStock | lsannes

First edition published 2023
by CRC Press
6000 Broken Sound Parkway NW, Suite 300, Boca Raton, FL 33487-2742

and by CRC Press
4 Park Square, Milton Park, Abingdon, Oxon, OX14 4RN

CRC Press is an imprint of Taylor & Francis Group, LLC

ISBN: 978-1-032-43798-9 (hbk)
ISBN: 978-1-032-43974-7 (pbk)
ISBN: 978-1-003-36967-7 (ebk)

DOI: 10.1201/9781003369677

Typeset in Times
by KnowledgeWorks Global Ltd.

For Lieutenant Thomas J. LeBel, USAF

Contents

Preface

I started flying airplanes in 1971, at Palm Beach International Airport. I was fourteen years old. My first lesson was on January 30, the day before Apollo 14 lifted off from the Kennedy Space Center. I know that because the day after my best friend and I had taken our first flying lessons, we drove up the coast of Florida with our fathers to watch the launch. In two days' time, I was pretty sure I had arrived in the space age.

I am now retired, after forty-one years as a commercial airline pilot. Between that first day and today, I logged something north of 18,000 hours of flight time. As with almost any retired pilot, I could write a book about decades of truly amazing adventure and experience, flying all over the world. I wouldn't trade it for anything; not the airplanes, from Convair 240 to the L-1011, 727, MD-80, 757, and 737 Max, not the layovers, from Dayton to Paris, Cairo, Lima, and Barbados, and most certainly not the colorful, talented, and wonderfully eccentric characters that I flew with. But that is not this book.

On a single, radiant New England summer day, my career … my life … took an abrupt, brutal, and wrenchingly painful turn. On June 25, 1978, I crashed a Grumman Yankee trainer in the mountains just north of Lake Winnipesaukee, New Hampshire. The man with me was a close friend from high school and college; he had flown with me dozens of times during our college years. He had recently graduated and been commissioned in the US Air Force. At the end of the summer, he was destined for pilot training at Williams Air Force Base.

He was in the left seat. As a current and practicing flight instructor, I was giving him some basic instruction, mainly just because he was a very talented and capable individual with an amazing future ahead of him. We flew a practice approach to Moultonborough, then climbed to the north, turning east along one of my favorite ground reference paths across the hills to the north of the lake. We flew just a few hundred feet above the terrain, following the contours over to a point where a right turn southwest-bound would take us across a saddle in the rim on the north shore of the lake, back out across the water. I had done this many times.

At some point, something went very wrong. It seemed apparent that we would not clear the next ridge. I took control, added full power, and established a climb attitude. This did not seem to help. The terrain sloped down to the left, but a left turn would have taken us into the ridge before we could complete it. I rolled into a longer right turn, away from the ridge, intending to come all the way back around to the northwest, pointing downslope toward a different escape route. The last thing I remember was expressing profound shock that we were not going to make it.

Two days later, a US Air Force Huey crew extracted me from the forest. I was in very bad shape, with both ankles shattered and a crushed vertebra, along with a broken collarbone and numerous lacerations, as well as being seriously dehydrated. When the first PJ reached the ground, I was several yards away from the wreckage, intent on dragging myself to a stream nearby. I told the PJ that I thought my friend might still be alive and to go to him first. In an unforgettably compassionate act of honoring my dignity, given that I was pretty close to expiring myself, he complied. He returned shortly and carefully told me that my friend was gone. I had expected that, and of course, since that young airman had already been at the main wreckage, he already knew it. But hope had gotten me this far, and hope, however forlorn, must never be surrendered. My young rescuer chose to let my hope adjust slowly, rather than crush it.

The Safety Board, as far as I know, never went to the site but examined the wreckage after it had been airlifted out of the woods. They did visit the hospital and take my statement, and eventually decided that I had "failed to obtain/maintain airspeed", made "improper in-flight decisions or planning", and "continued flight over rugged terrain". They also mentioned downdrafts as a weather factor, probably because I told them that's what I thought had happened. They even threw in the entire section on mountain flying from the pilot's operating handbook and added a statement that when the conditions were "right" in the area we were flying in, it was extremely turbulent with downdraft and heavy wind shear effect.

The problem with all of that is that it had been a beautiful summer evening. There was no significant wind at the surface and none aloft to speak of. The air had been smooth. In fact, the reason we flew the route we did is because it was such a perfect evening to do it. In the margin of my copy of the Board's report, my father scribbled, "no indication of winds like this in the forecast, so what is the point I filling up this report with reprint?" Today, I still have no good idea why we were sinking in the first place. Or if we even were. I've looked at the weather data a thousand times and still don't see where a downdraft sufficient to exceed to airplane's climb capability would have come from. I'll never know whether we could have successfully turned left; I didn't think so at the time. I'll never know if we actually would have cleared the ridge; I didn't think we would. I'm pretty sure I got us into an accelerated stall during the turn, but I have no recollection of excessive bank. I will never know if a few degrees less of bank would have made a difference.

I realized a long time ago that I would never know what had happened.

There wasn't enough of an investigation, yet in fact, the best investigation possible may just as well have found no more than this one. There simply was not enough factual evidence to work with. The phrases included in the accident report are nothing more than selections from a pull-down menu of factors designed to bend and fold the accident into a preconceived taxonomy. People don't like uncertainty. They need a reason; they need to be able to organize events in their mind. The disconnect between what the actual weather was and all of the boilerplate pasted into the report unveils the true nature of our relationship with accidents. While phrases like "failed to obtain/maintain airspeed" can neatly pigeonhole an event and, by the way, create some distance between ourselves and the hapless pilots who failed to obtain/maintain airspeed, they obfuscate and erase for history the complicated, even byzantine web of nuance, variation, and complexity that creates a singularity of conditions for each and every accident. You have to wonder how many accident reports containing the phrase "failed to obtain/maintain airspeed" are similarly uncertain.

The one inescapable truth that I do know is that what happened was my responsibility to prevent.

So, I have spent the last forty years trying to understand more about the nature of accidents and to learn all I could about how to better execute that responsibility. That journey is what this book is about. I'm not going to follow a traditional path in describing this exploration. During my career, I have been involved in a number of major airline accident investigations, from AVAir 3378 to TWA 800 and several in between. I have participated in a number of safety initiatives, from the One Level of Safety effort in the early 90s that brought us Part 119 to the Commercial Airplane Certification Process Study in 2001. I have been a member of several rulemaking committees, most notably the years-long Ice Protection Harmonization Working Group. I had the privilege of being a part of building an entirely new and original airline. Along the way, I have attended more meetings, read more papers, and sat through more safety seminars and recurrent trainings than I really want to remember.

Most of that effort ... not all, but most of it ... has been very productive and meaningful. I think we've done a lot of good things; a lot of people have certainly worked very hard to do so. But none of it has really answered the mail, so-to-speak, from my perspective as an accident survivor, because none of it has ever really come to terms with the most fundamental truth about any accident. That truth is simply that you will never, ever, not in a million years, ever see the one coming that gets *you*.

We are hamstrung, to a great extent, by the character of human cognition. We generally think of accidents in one of the two ways: we describe them through a story or we describe them as a statistic. Both of these approaches are powerful and useful tools, but both are frozen in time because they can only describe what has already happened. Any story is colored in by the language used and the meaning of that language. Words like "continued" are fairly black-and-white, describing a simple fact, but they can also imply additional meaning depending on context. A word like "failed" is also quite factual, but somewhat more prone to pejorative interpretation. But a word such as "improper" is designed to convert opinion into fact through unabashed pejorative judgment.

Furthermore, a story is built around a linear model of cause-and-effect. The effect is usually quite clear; the cause is more elusive. One way to backfill a story, beginning with the effect, and moving toward the cause, is through the aforementioned choices of language and the meaning therein. This gives the appearance of accounting for all of the detail and nuance, which promotes a sense of certainty. As history, the story can be deeply flawed for all of the reasons I have described. But the real problem with applying a story model to the future is that there is no story about an accident that has yet to occur because it hasn't actually happened. So, we are stuck with thousands of aimless causes and millions of meaningless details that cannot be linked to an effect because there is no effect. Not yet. We can guess, and if we guess well, there will be no accident. On the other hand, if we don't capture the nuance, don't see the patterns, then our guess, if we even think to make a guess, will miss the mark.

We can and do link causes and details statistically, but a statistical model is built from many, many stories, and a probability is created through a pretty binary, black-and-white assignment of value ... pull-down menus ... derived from the language we use to describe the stories. It is only as good as the stories, and the language, that it is built from. Even if we assume those stories to be watertight, we are still trapped by the very nature of a probability. In the end, all probabilities have a value of less than one, meaning that each data point represents a probability of something happening while simultaneously representing a probability of something not happening. None of them ... none ... have ever leapt out from under a bell curve and actually happened. This problem approaches the thought experiment of Schrodinger's cat, in which a cat is considered to be simultaneously alive and dead because the probability of a subatomic particle decaying and causing a poison to be released in the next hour is 50/50. When something does occur, when the particle does decay and release the poison, it suddenly has a probability of one, meaning that it is no longer a probability. And it is too late.

Statistical analysis is intended to quantify uncertainty, not to be used as a decision tool in the face of actual uncertainty. This conundrum demands that we think differently. Instead of anticipating certainty derived from historical models, we need to anticipate uncertainty, and in particular, a condition that I refer to as persistent vulnerability. There is not a single pilot involved in an accident who gets up that morning and sees someone in the mirror who, later in the day, will kill several hundred people. Not a single one of us can look in the mirror on any given morning and be certain.

So, while that is the last time I'll mention Schrodinger and his famous cat, I'm going to take a somewhat different path from that conventionally used. This book is a pretty deep dive into how we might think differently. I'm going to bring in a lot of thinking, some contemporary, some quite ancient, and hopefully show how it applies to what we actually do. The unwritten message that lies at the foot of human factors in aviation is that we are all human beings first and aviators second. Flying is a very singular, unique human endeavor, and we are not born to it nor can we be trained to it with any expectation of perfection. We must approach it not as procedurally programmed automatons but as human beings vested with the most profound responsibility in all of human endeavor.

Before launching into the text, a serious question might be asked: why? In the last several decades, commercial aviation safety has improved by whole orders of magnitude. Indeed, the first and most important consideration is that we must do no harm. We work in a well-designed, well-operated system that has yielded tremendous success. That success cannot be jeopardized.

Nevertheless, commercial aviation stands upon a precipice. The demand for new pilots is relentless. The path toward becoming a pilot has narrowed and become cost-prohibitive. The opportunities for gaining experience have diminished severely. At exactly the same time, the operational world of the commercial aviator has become vastly more digitized and nearly cybernetic. Such digitization is a powerful tool that has brought about significant improvement in standardization, consistent performance, and repeatability.

But human beings cannot be digitized. More importantly, we fly in an analog sky, a sky that really could not care less about binary code. There are analog lessons to be learned. My aim is to pass along as much of the analog world as I can, in the hope that new aviators, and some not-so-new aviators, can integrate it into the digital world of their future.

About the Author

Steven D. Green started flying at age fourteen and soloed on his sixteenth birthday in 1972. He began his airline career flying a Convair 240 for Providence Airlines around the Great Lakes, then flew Fairchild Metroliners up and down the east coast through the 1980s and then all over the world for TWA, Eos Airlines, and American. He holds a B.S. in Aviation from Louisiana Tech University. He also holds an Airline Transport Pilot license from the FAA with type ratings for the Boeing 767, 757, 737, MD-80, EMB-145, and the Fairchild SA-227, as well as a Flight Engineer license with a turbojet limitation. He flew professionally for forty-one years, operating at all levels of the industry including Part 135 commuter operations, Part 121 supplemental operations, and Part 121 flag operations. Beginning in 1986, he has participated in numerous aircraft accident inquiries and investigations as a representative of the ALPA including TWA 800, TWA 843, Business Express N811BA, Simmons 4184, Comair 3272, AVAir 3378, and AVAir 3464. He received his formal accident investigation training from the Transportation Safety Institute in Oklahoma City in 1993. For two years, he taught two segments of the ALPA's Basic Accident Investigator Course. He also participated in the FAA Commercial Airplane Certification Process Study in 2001–2002 and contributed to the drafting of the final report. Association with the 1994 Roselawn accident involving Simmons 4184 led to work with ALPA's Inflight Icing Certification Project, which included participation on the Ice Protection Harmonization Working Group ARAC. Since 1994, he has been actively involved with icing issues and has written a number of papers on the topic and delivered several oral presentations. Since 2004, he has actively consulted for both National Aeronautics and Space Administration and the FAA on the subject of airframe icing. He and his wife have lived in Vermont for thirty-three years and have two grown sons. He retired from American Airlines in 2021 as a Boeing 737 captain.

List of Abbreviations

AC alternating current
ACARS Aircraft Communications Addressing and Reporting System
ADF automatic direction finder
ADIRU air data inertial reference unit
ADR air data reference
AGL above ground level
AIM Airman's Information Manual
ALPA Air Line Pilots Association
AoA angle of attack
APU auxiliary power unit
ARAC Aviation Rulemaking Advisory Committee
ARFF aircraft rescue and fire fighting
ARINC Aeronautical Radio, Inc.
ATC air traffic control
ATIS automatic terminal information service
BTCU borderline tolerated conditions of use
CAA Civil Aeronautics Authority
CAVU ceiling and visibility unlimited
CB circuit breaker
CFR crash fire rescue
CRM crew resource management
CRT cathode ray tube
CSD constant-speed drive
CVR cockpit voice recorder
DME distance measuring equipment
EASA European Union Aviation Safety Agency
ECAM electronic centralized aircraft monitor
EFIS Electronic Flight Instrument System
EGPWS Enhanced Ground Proximity Warning System
EICAS Engine Indicating and Crew Alerting System
ELT emergency locator transmitter
ETOPS Extended-range Twin-engine Operations Performance Standards
FAA Federal Aviation Administration
FAR Federal Aviation Regulation
FCPC flight control primary computer
FMC flight management computer
FMEA failure modes and effects analysis
FMS Flight Management System
FPM feet per minute
GPWS Ground Proximity Warning System
HF high frequency
HRO high-reliability organization

ICAO	International Civil Aviation Organization
IAF	initial approach fix
ILS	Instrument Landing System
LOFT	Line Oriented Flight Training
LORAN	long range navigation
MAP	missed approach point
MDA	minimum descent altitude
MEL	minimum equipment list
METAR	meteorological aerodrome report
MORA	minimum off-route altitude
NDB	non-directional beacon
NOTAM	notice to airmen
NTSB	National Transportation Safety Board
PA	public address
PBN	performance-based navigation
PFD	primary flight display
PJ	para-rescue jumper
PRA	Probailistic Risk Assessment
QRH	Quick Reference Handbook
RNAV	area navigation
RVR	runway visual range
RVSM	reduced vertical separation minimums
SAL	Saharan Air Layer
SAFO	Safety Alert for Operators
SB	service bulletin
SFAR	Special Federal Aviation Regulation
SOPs	standard operating procedures
SSA	System Safety Assessment
TAF	Terminal Area Forecast
TALPA	Takeoff and Landing Performance Assessment
TCAS	Traffic Collision Avoidance System
TR	transformer-rectifier
TWA	Trans World Airlines
UPS	United Parcel Service
UTC	Coordinated Universal Time
VHF	very high frequency
VNAV	vertical navigation
VOR	VHF omni range

Acknowledgments

I would like to thank Captain Shem Malmquist, FedEx, First Officer Danielle Gibault, UPS, Captain Rob Buck, Delta (retired), and John Zimmerman, editor at Air Facts Journal, for their encouragement and review of this book.

I would also like to remember the late Captain David J. Haase, TWA (retired), for his kind support and encouragement to get these ideas down on paper. Dave was an exceptional airman, one of the best I ever flew with. He opened many doors for me throughout my career in air safety.

1 The Nature of Accidents

THE TENTH AXIOM

Among my collection of antiquarian aviation books and manuals, I have a copy of *Civil Aeronautics Bulletin No. 5, Flight Instructor's Manual*, published in June 1939. At the end of the manual are listed ten axioms for the pilot. My personal favorite is the tenth axiom, which says that "The capable and competent pilot will never allow an airplane to crack up …".[1] It then goes on to state that if a crash is inevitable, the pilot (apparently no longer classified as capable or competent) should maneuver the aircraft "in a manner to ensure that no injuries will result to himself or his passengers",[2] which, according to the manual, is almost always possible.

Most of us in the business of flying airplanes easily recognize the narrow-minded, myopic approach to safety that is captured here, and the historically aware might also recognize a Victorian tone that dates the original construction of this axiom back to World War I or slightly earlier. We could easily dismiss it as quaint. But I think that too quick a dismissal underestimates the impact that this axiom has had on the history of our own work in safety and the paradigms that go with that work. Rather, a careful unpacking of the axiom is worth taking some time to do.

There are probably some pretty deep cultural undercurrents at work in this axiom. It is hard to overlook that tone, which subtly flavors the accident as a fall from grace, suggesting that whether an airplane crashes or not is largely a matter of free will. Indeed, I suspect the axiom is rooted in the broader social and religious thinking of the era, which tended to emphasize the role of free will in overcoming inherent depravity. At the time, this thinking was being steadily challenged by accelerating technology, the science that lays behind that technology, and the increasingly irrefutable role of determinism in that science. The frequent failures of early technology presented powerful emotional trials to society … witness the crash of TWA 599 on March 31, 1931, which cost the life of Knute Rockne, among others … particularly since the deterministic aspect to causality made the control of that causality more and more complex.

Yet the accident reports of the same era do not consistently reflect the tenth axiom at all. In the Rockne crash, the investigation centered on poor glue application during the lamination process of a wooden wing spar and steered away from questions regarding the pilot's decision to depart, although that decision was debatable.[3] In other accident reports, there are as many references to poor judgment as there are to newly understood technical phenomena, such as icing, radio range failures, and weather conditions seriously different from those forecast. Indeed, there are even some rather fantastic technical explanations … one of my favorites was the DC-3 crash of Pennsylvania Central Airlines Flight 19 near Lovettsville, Virginia in 1940, in which one of the hypotheses suggested that the concussion resulting from a nearby lightning strike had "smashed in" the windshields.[4]

As it happened, a few years before my edition of the Flight Instructor's Manual came out, Jimmy Doolittle managed to destroy a brand-new Lockheed Vega after he

DOI: 10.1201/9781003369677-1

had loaded it up with his family and all their worldly belongings and then attempted to take off, overweight, from the rutted frozen mud of Mitchell Field. Not too long afterward, he snapped the ailerons off of a highly modified Travel Air while tooling down the runway with more airspeed than altitude … the modifications were not as well thought out as they might have been … leading to yet another successful parachute deployment. Of course, both of these events took place sometime after his first parachute jump, made after he pulled the wings off of a Curtiss P-1 Hawk fighter when attempting an outside loop, a maneuver coincidentally prohibited by his commanding officer.[5] Although he himself made no excuses about these events in his autobiography, at the time he was considered rather capable and competent, having, after all, just demonstrated flight solely by reference to instruments. He had to have been something like that when he released the brakes on the lead B-25 off the deck of the Hornet.

Within two years of its publication in my copy of the manual, the tenth axiom was severely tested in the cauldron of World War II. Thousands of pilots, some perhaps marginally capable or even minimally competent, lost their lives making a somewhat more profound moral argument than one advocating the role of free will in preventing airplanes from "cracking up". In the meantime, on the home front, extensive research was being done to investigate such topics as pilot skill, judgment, and emotional control. In 1939, the Civil Aeronautics Authority was examining the relationship between breathing patterns and personality, with a view toward identifying the more extroverted candidates, characterized by one researcher as possessing "such features in their daily lives as simultaneous interests in a number of business enterprises, frequent entertainment, lack of interest in abstract thought, hearty dispositions and carefree attitudes",[6] a remarkably prescient description of an airline pilot if ever there was one. By 1943, the University of Rochester was researching "The Ability to Take It",[7] which has always seemed to me a bit of a Bogartian title for a research paper, but nonetheless consisted of evaluating several different techniques for measuring human resistance to pain and fatigue, among them a continuous electrical shock of increasing voltage or having a wedge slowly clamped down between one's knuckles. Eventually, the shock technique, along with a procedure for measuring how long a man could hold 60% of his maximum grip strength, was recommended for further study. The knuckle wedge idea was dropped.

A serious question might have been asked, and apparently was asked at some point, about how many of the fellows who were actually "taking it" in the skies over Europe and the Pacific at the time of this research would have passed the shock test. I suppose the grip test might have been useful for trying to figure out whether a pilot candidate would be able to hold onto the cockpit window frame while reaching back into a disintegrating B-17 to grab the parachute that he had overlooked when climbing out, as was claimed in at least one instance. No doubt the breathing patterns of pilots who were being shot at could be of interest, particularly the "unconscious vocalizations" that CAA researchers had earlier associated with the more "extroverted" research subjects.[8] I'd be willing to bet that some of the more introspective combat pilots, previously thought to be inclined toward a quieter form of "visual phantasy thinking",[9] may have exhibited a few unconscious vocalizations as well. In any event, these kinds of studies characterize the type of the shot-in-the-dark research done during the pre-war and war periods toward methods of identifying that select group of human beings who were more naturally inclined to never allow an airplane to crack up.

In his book *The Psychology of Flight*,[10] published in 1950, Alex Varney goes into exhaustive detail about the disastrous effects of unchecked emotion, describing a bevy of behaviors that, when read in a contemporary context, seem far more likely to appear in illegal street drag racing than aviation. I don't think he seriously believed that a pilot attempting to fly the Atlantic would actually start off in the opposite direction in order to buzz the blonde working in the cigar factory and then descend halfway across the ocean to circle a couple of icebergs, consequently running out of gas short of the destination. Nonetheless, examples of character weakness such as these, and his emphasis on the proper use of free will to overcome such weakness, continue to resemble an echo of that Victorian-era notion of depravity, perhaps now lightly salted with some of the Freudian fascination with the subconscious popular during the first half of the 20th century. Varney even revisits the ever-recurring notion of accident-proneness, a subset of ineptness containing those folks who, by virtue of clumsiness, absent-mindedness, and a general inability to navigate the more Hobbesian aspects of the day, are predicted to have regular accidents. On the other hand, he makes the case that a strong character is reflected by the disciplined control of emotional response and the "training" of "reflexes" so that the aerodynamically correct action is taken without thought when faced with debacle. From such ideas appeared iconic images of the era such as the comic strip hero Steve Canyon.

So, the tenth axiom persevered, appearing as late as 1958 in the Pilot Instruction Manual, having been slightly enhanced to state that "A capable and competent pilot will never allow an airplane to crack up out of control".[11] By this time, industrial process management, developing from roots in Frederic Taylor's planning rooms and time-motion studies, had begun to offer the certainty of mathematics to the management of human behavior within organizations. Herbert Simon, building ideas that would shortly introduce the notion of artificial intelligence, believed that human thought could be characterized by mathematically predictable patterns of information processing. Shortly after World War II, Simon had argued that

> Two persons, given the same skills, the same objectives and values, the same knowledge and information, can rationally decide only upon the same course of action. Hence, administrative theory must be interested in the factors that will determine with what skills, values, and knowledge the organization member undertakes his work.[12]

In other words, a flowchart could be drawn showing how a person would think in any particular situation and human performance could be predicted and modified deterministically, in much the same way as the operation of a machine might be. With this in mind, the capable and competent aviator could, in theory, be modeled. By carefully managing the aviator's skills and particularly by managing his knowledge and information, his decisions could be predicted and controlled. If accidents could not be eliminated through the autonomous pilot's proper application of free will, perhaps they could be eliminated through management systems that more or less dismissed the notion of free will altogether.

Just a few years after the improved 1958 edition, Chuck Yeager managed to "crack up" the Lockheed NF-104 at Edwards Air Force Base, dropping out of 100,000 plus feet about as out of control as could possibly be achieved. Perhaps a little too much free will pointed in one direction and not enough pointed in the other, and definitely a case of pretty skinny flight safety assessment; nevertheless, it would be hard to describe Yeager as anything but capable and competent. On the other hand, it is probably safe to say that Herbert Simon could have spent the rest of his life trying to model Yeager's thought processes.

Which brings us to the beginning of the 21st century, in which James Reason, among many others, has argued tirelessly that human error is a universal condition. He has distinguished between the "person" approach to safety, which single out the errors of individuals, placing blame on them for mindlessness, ineptitude, or moral weakness, and the "system" approach to safety, which instead focuses on the environment within which individuals work and strives to create defenses that capture or prevent error and/or mitigate the effects of error.[13] In this latter approach, he disrupts Simon's notion of consistent human behavior, departing from the attempt to create that predictable, repeatable working environment and instead striving to build a resilient environment which is capable of tolerating inadvertent deviation. In his 2000 essay for the British Medical Journal, *Human Error: Models and Management*, Reason explained that in particular, high-reliability organizations are persistently engaged with the possibility of failure,[14] which is the first of the five characteristics of a high-reliability organization cited by Karl Weick and Kathleen Sutcliffe in their pathfinding work on the subject.[15]

Yet today, we still borrow freely from the tenth axiom and some of the ideas Varney presented when we use terms such as "discipline" and "excellence" in opposition to words like "complacency" and "apathy". On the other hand, we borrow from Simon and administrative theory as we simultaneously create structures designed to manage error, risk, and culture. Like many disciplines, we have become engaged, perhaps even a bit enthralled, with "big data", and we struggle to understand the single catastrophic outlier that escapes statistical prediction. Events such as Asiana at San Francisco, or the Dreamlifter landing at the wrong airport in Wichita, get caught in the whirlpool between the notions of a capable and competent pilot as described by the Flight Instructor's Manual and the contemporary systems management ideas that have evolved from Taylor, through Simon, and into the world of big data.

The whirlpool persists in part because the cultural argument over the primacy of either free will or determinism just won't go away, no matter how progressive our analysis. The argument is nearly as old as dirt, and it remains hotly relevant today as neuroscience and other disciplines explain more and more of human behavior from a deterministic point of view, rendering the search for the source of free will even more problematic (though hardly pointless). The argument inevitably stalls on the issue of moral responsibility, and it is here that we debate the role of the system, complexity, and organizational behavior as explanatory in contrast to individual incompetence, willful violations, or the "bad apple" described by Sidney Dekker.[16] To veer too far toward a fully deterministic interpretation is to risk doing away with personal and moral responsibility altogether, dispersing into the so-called blunt end on a Simon-esque mission to preset the variables to fail-safe values, while inadvertently begging the question of where, along the continuum of systemic determinism, anyone will apply the free will required to effect the management of anything. Reason himself, in later work, has argued that the pendulum may have swung too far[17] and that a re-examination of the role of those people at the "sharp end" may be in order.

It is indeed hard to remember, after so many years of administrative theory and systems management, that our capable and competent aviator actually remains a fully autonomous actor in the whole scheme of things. The regulatory obligation is, and always has been, that the pilot is the final authority as to the operation of the aircraft and command of the crew, no matter how much a whole cast of corporate characters would like to water him down. That authority presupposes a role for free will and transcends the conditions of employment, making the pilot-employee something of a greased pig for all styles of management.

But the whirlpool may also persist because of the structure of thought itself, particularly the ways in which we are constrained to think about accidents as distinguished from how they actually occur. Part of that constraint is obscured by the clarity of hindsight, which we regularly recognize but then just as regularly marginalize, on the notion that it is a previously stipulated, universally recognized condition … "hindsight is twenty-twenty" … or words to that effect. But the impact of hindsight on our interpretation can scarcely be overstated, for the simple reason that, in all cases, you cannot think of a thought until you think of it. This is a genuine conundrum, conditionally but nonetheless thoroughly overwritten by hindsight. We then tend to marginalize this idea as well, while routinely attempting to out-flank it, under the nearly unconscious assumption that the particular thought must actually exist somewhere and is simply not being acquired, usually through a lack of diligence, foresight, or will. Sidney Dekker refers to this as a

> … naive Newtonian scientism: total knowledge of the world is achievable; the world is "out there" as an object entirely separable from observers. People will know the truth if they are fully rational, once the correspondence between the picture in their mind and reality in the world is perfect.[18]

THE MODES OF HUMAN THOUGHT

The psychologist Jerome Bruner, in his book *Actual Minds, Possible Worlds*, describes what he believes are the two principal modes of human thought: the narrative and the paradigmatic. The paradigmatic mode is a rigorous, mathematical description and explanation.[19] It is the language of logical argument. The narrative mode, on the other hand, describes human intention and action, including the changes in circumstance, modifications, and consequences of such actions.[20] A good narrative begins with the ordinary, the routine, and then features a sudden reversal of circumstances, often referred to in literature as a peripeteia. The peripeteia is the occurrence or development that disrupts the ordinary and puts it at risk. In his paper, *Culture and Human Development: A New Look*, Bruner argues that "These narratives typically depict a canonical state of things and a deviation from that state. Stories are means for making these deviations comprehensible, if not acceptable".[21]

Bruner makes the case that a story has a skeleton, shown as a sketch in Figure 1.1. It begins with the canonical, ordinary state of things, experiences the peripeteia, and

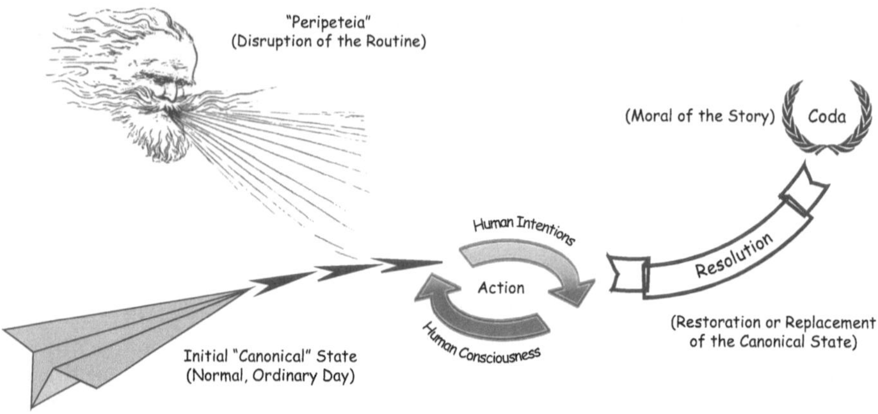

FIGURE 1.1 The narrative skeleton, the way stories are told.

this is followed by an action or series of actions, an attempt to undo the peripeteia. The action is followed by a resolution, either the restoration of the canonical state or its replacement by another canonical state. Finally, the story contains a coda or a discourse in the lessons learned. Returning to *Culture and Human Development*, Bruner says that "narrative seeks to render the ordinary as if it were not only majoritarian but also obligatory, whereas the ultra-ordinary is made to seem optional and subject to choice".[22]

Herein lies the actual constraint on how we think about accidents. The management of risk, error, and safety culture, essentially the entire system approach to safety, is organized and executed within the paradigmatic mode of thought, as illustrated in Figure 1.2. In the accident investigation discipline, we make a dedicated effort to produce a paradigmatic description of the accident, in order to identify and control the variables that have conspired to align the holes in Reason's cheese. We commit a lot of ink to investigations into oversight, culture, training, and even more specific aspects such as operational control and the approval of supplemental-type certificates. Nevertheless, one way or another, we always make time to stop and visit issues of character, as reflected by the flight crew's behavior. It is virtually impossible to separate the details of the accident paradigmatically from the human intentions and actions and the resulting changes in circumstance, modifications, and consequences that result. There is a story there, waiting to be told. The accident is clearly differentiated from routine operations by a peripeteia, a deviation from a canonical state. Our capable and competent aviator is a protagonist, or perhaps an antagonist, depending on one's point of view, in a story that is inescapably interpreted through the narrative mode of thought. There will be a resolution, either a recovery from the disrupted flight or a somber change in procedures or training resulting from what was learned. And there is always a coda, a moral of the story, usually embedded in the probable cause statement.

Both the paradigmatic and the narrative modes of thought work to create certainty where none exists, either a calculated, repeatable certainty or a certainty derived from comprehensibility, from knowing the whole story from beginning to end.

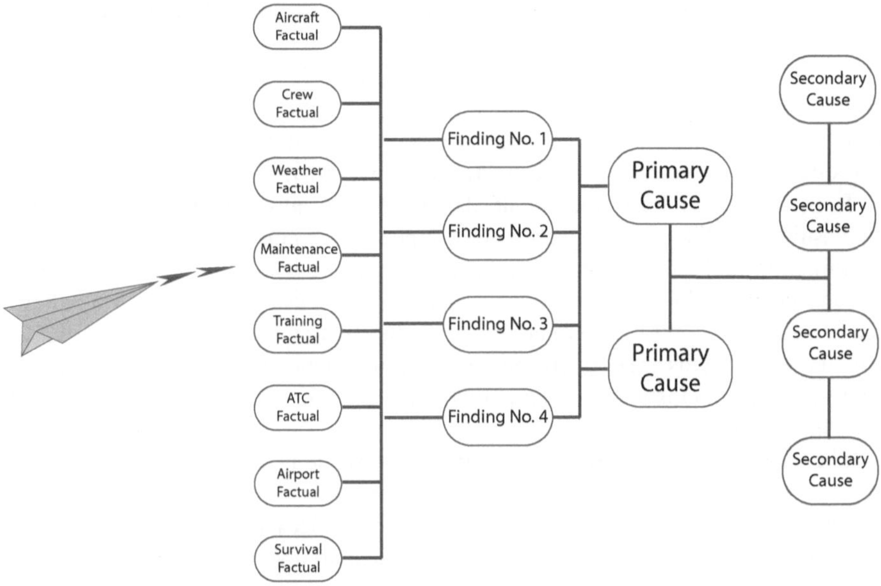

FIGURE 1.2 The paradigmatic approach to an accident report.

We have a strong need to trust the day … the sun comes up, the sun goes down … and the overwhelming sense of deceit experienced through proximity to a serious accident or its immediate aftermath is demoralizing, heartbreaking, and completely fractures any trust in the day that we might have had at breakfast. Such feelings create a powerful incentive to put as much emotional distance between ourselves and the breakdown of certainty as we possibly can. One way of doing that is to isolate the event, place it under a glass case, and stare at it from the outside, while developing a narrative structured to remove the sense of deceit, the apparent unpredictability, as well as the irrevocable impossibility of a second chance, and replace them with a mechanical, structured comprehensibility that offers the hope of a sort of do-over, as in I'll never let that happen to me, thereby reinstating trust in the day.

But Bruner goes on to point out that authentic narrative also contains two different landscapes: the landscape of action, in which the participants are the origin of actions, intentions, and goals, and the landscape of consciousness, which describes the participant's thoughts, knowledge, and feelings, or perhaps more importantly, what they do not think, know, or feel.[23] Consciousness does not lend itself easily to paradigmatic investigation; thus, an accident report is typically pretty long on the landscape of action and somewhat short on the landscape of consciousness. The initial story will usually be rather flat and may not do much to assuage our need for comprehensibility. What we can discern about the participant's state of consciousness may be incomplete, truncated, or worse but still offer absolutely no precursory perceptions that might serve to warn of what we already know about the rest of the accident.

Where was the flaw in the knowledge or thought processes that led to an undesirable aircraft state? Surely there must have been a flaw. What kinds of signposts in those thought processes should we be looking for that will forewarn of such a catastrophic outcome? We'll never know of course, since the crew have ended up, in the memorable words of the fictional Squadron Leader Colin Harvey from the movie "Battle of Britain", "spread from one end of this field to the other like strawberry jam" … and to make matters worse, they foolishly failed to vocalize their entire day's thinking for posterity on the cockpit voice recorder.

But since the tenth axiom, and similar tropes, can introduce an element of presupposition into the narrative, voids in the landscape of consciousness can easily be backfilled to make the narrative work, to make the "deviations comprehensible, if not acceptable".[24] We backfill the landscape of consciousness with what we believe the consciousness must have been and what the participants must have known or felt, and these beliefs are constructed almost entirely from our knowledge of the outcome and from an instinctual suspicion that the crew, in one sense or another, must have fallen from grace.

The subtle, quiet effect of backfilling the landscape of consciousness is that we never doubt our own grip on comprehensibility. We are not in the least bit inclined to consider possibilities that remain incomprehensible, at least to the extent of the available knowledge. It is as though we expect to be able to invert the narrative, casually walk around to the other end of the lens through which we view the past, and look forward into the future, fully anticipating that the lens works both ways. It doesn't; it never has. The morning of the day of the accident looks the same as any other morning; the evening looks like no other evening at all.

THE HIDDEN AXIOM

In 1978, I was working as a flight instructor in central New Hampshire. On a warm, cloudless, and quite calm summer evening, a friend of many years and I were flying a Grumman trainer a few hundred feet above the terrain to the north of Lake Winnipesaukee, having just flown a practice approach at Moultonborough. We were following a route that would take us to the east of the hills on the north shore of the lake and point us back toward Laconia.

At some point, something went terribly wrong. Despite the absence of any surface winds or forecast winds aloft, I had a strong sensation that we were starting to sink and it did not appear that we would clear the next ridge. I took control of the airplane, applied full power, and established a climb attitude. This did not appear to help the situation, and I began a long turn to the right, upslope initially but away from the most proximate terrain, with the intention of turning approximately one hundred eighty degrees to a downslope route.

I was not successful. The airplane clipped the trees and came to rest deep in the forest, hardly breaking a branch as it fell. The following evening, the ELT was picked up by a passing SAS DC-8, and the next afternoon, after about forty-three hours, we were located by a USAF UH-1 Huey, which dropped in a couple of PJs to initiate a rescue. By that time, my friend had passed away. I was in rather serious condition, with shattered ankles and a crushed vertebra as well as serious lacerations, a broken collarbone, and a collapsed lung … but, thanks to something like six or seven combined tours dropping out of similar helicopters into the Vietnamese jungle, I was successfully packaged up, lifted through the forest canopy into the old Huey, and flown to the hospital to begin a lengthy recovery.

To this day, over forty years later, I still do not know what actually happened. My perception was that we had encountered a downdraft, as I had experienced that before, but I have never understood how that energetic a downdraft would have occurred on such a calm summer afternoon. I suspect that, in the turn, I managed to induce an accelerated stall. The official investigation did not bother with any of those questions, or tear down the engine, or even visit the crash site. They were content with my hospital bed interview and the popular accident report refrain of "failed to maintain flying speed". No kidding.

The narrative mode of thought is dependent on the landscape of consciousness, but the landscape of consciousness will not actually respond to naive Newtonian scientism. Backfill after the fact as we may, the entire frame of thought is hostage to the hidden axiom, so to speak, the axiom that more or less plays the joker in the deck of axioms, trumping all the rest. That axiom says that you will never, ever, not even in a million years, see the one coming that gets you.

Of course you won't. If you saw it coming, you wouldn't let it get you.

This is a pretty unsettling proposition; it has unsettled me for decades. It is a corollary to the idea that you cannot think of a thought until you think of it. The accident cannot be experienced in the narrative mode of thought because actual comprehensibility is not possible until the narrative is complete, at which point it is obviously too late. The only way we can interpret experience in real time is by correlating a developing sequence of events with one or more known narratives, in order to identify a likely comprehensibility, but to do this, we must still make an educated guess as to the outcome. The probability of accurately guessing the outcome improves with experience and education, but it will always require us to make the correct correlation with known narratives, and it will always require us to assume a conclusion not yet in evidence. Acting on that conclusion before it has manifested requires an acceptance of persistent vulnerability, the opposite of certainty, and persistent vulnerability has no place in either the tenth axiom or administrative behavior … or, for that matter, to anyone's trust in the day.

Yet the term "will never" in the tenth axiom is an absolute term, addressing all which has not yet happened, leaving no room whatsoever for errors of perception, presumptively requiring that the aviator, by virtue of being capable and competent, will always see the accident coming, as if he could just eyeball a line of holes right through Reason's cheese. This is the net effect of a lifetime of trust in hindsight of an entire body of experience built from comprehensibility. In fact, the aviator who actually has the accident, regardless of whether he is capable or competent, is never going to know what hit him.

Herein lies the fundamental flaw contained within the tenth axiom. By creating the impression that the capable and competent pilot will always see the crack-up coming, it also promotes the contrapositive idea. The contrapositive says that, if you reside within the set of capable and competent pilots, then the threats that you see, that you become aware of and can identify, are the only threats that exist. This is precisely what Dekker means by naive Newtonian scientism. Total knowledge of the world is achievable … people will know the truth if they are fully rational … ergo, if you are fully rational, thus capable and competent, then you will see all possible threats.

Yet if the hidden axiom is true … if you really won't ever see the one coming that gets you … then why would that particular shortcoming be limited to the single threat that manifests into an accident? Is that the only threat that you won't see? Is it possible for threats to remain unseen while never manifesting into anything at all? If you won't see the one coming that gets you, is it also possible that you may not see dozens, hundreds, of "ones" coming that go right by you? And if that is true, how can our capable and competent aviator ever hope to apply free will in any way to manage such unseen threats?

A DIFFERENT WAY OF THINKING ABOUT ACCIDENT PREVENTION

One way to do so is by creating, updating, and executing a strategy that diverges from a focus on what goes wrong and instead concentrates on ensuring that things go right. The strategy begins with what we know works, remains proactive and anticipatory, and is constructed to tolerate variability in performance and adapt to modification in the operating environment. The key to tolerating performance variability and modifications to the environment is to protect the margins.

Margins are typically understood by pilots within the context of aircraft performance. We evaluate performance margins every day, from net vs. gross takeoff performance, through 1.3 g cruise altitude capability all the way to 167% of the landing distance … margins are our daily bread. Yet the concept of margins is much broader than aircraft performance. The most fundamental margins we protect on every flight are aerodynamic margins … angle of attack, bank angle, and g-loading … and altitude margins … minimum enroute altitudes, driftdown capability, grid MORAs, minimum crossing altitudes, and approach minimums. We protect margins of aircraft configuration and system functionality, such as single pack altitude limits and single generator diversion requirements. Figures 1.3 and 1.4 provide a couple of common examples of margins.

At a different level, we protect margins of human performance. Early in the postwar research done on human factors in aviation, Stanley Roscoe identified the concept of residual attention. He said that, "During routine flight operations a pilot's attention capacity exceeds the moment-to-moment demand by varying amounts …".[25] He then explained that "a blunder occurs because the perceptual, judgmental, and motor demands of the moment exceed a pilot's momentary attention capacity".[26] Protecting the margin of residual attention is critical and is precisely what rules pertaining to such things as a sterile cockpit are aiming at.

Rene Amalberti, the original head of human factors and flight safety at the European Joint Aviation Authorities, has pointed out that workers must execute their jobs within an envelope of possible choices and actions, and that the envelope is continuously influenced by much broader organizational and social forces and factors.[27] This idea is shown graphically in Figure 1.5. The capable and competent aviator of today is suspended somewhere between the Tayloristic, one-best-way prescribed procedures, a need for efficient, expeditious application of labor, and the struggle, in a proceduralized world, to retain an identity of craftsmanship, of artisanship really,

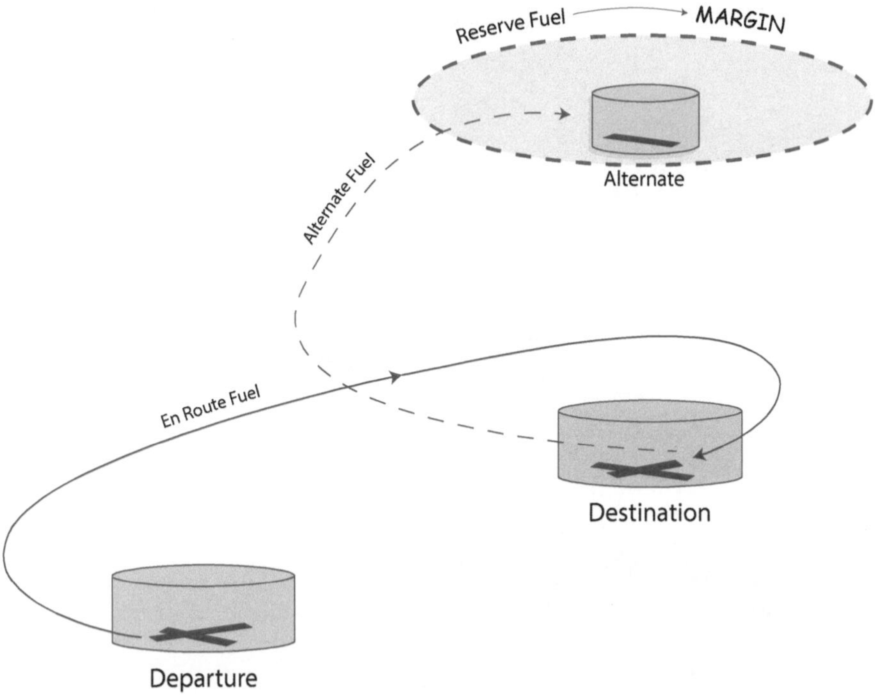

FIGURE 1.3 Reserve fuel is a margin.

through the application of free will. From a utilitarian perspective, it can be very easy to conflate economic efficiency and expeditious labor with craftsmanship, but this is misleading; the former is simply one of several outcomes resulting from the latter. Further, the "lack of interest in abstract thought, hearty dispositions and carefree attitudes"[28] noted by the CAA in 1939 can promote an almost Yeager-esque, right-stuff understatement of risk. The tension between these three ideas creates a drift, beginning at what Amalberti calls the "initial safe space of action", and migrating toward the "borderline tolerated conditions of use",[29] the absolute edge, a position which appears to optimize efficiency and economic benefit while retaining an acceptable level of risk, but beyond which there is no further margin, no safe space of operation at all. This natural drift is facilitated by the normalization of deviance. Conveniently, these days the drift toward borderline tolerated conditions of use is monitored by flight operations quality assurance (FOQA) programs … but a FOQA flag comes to this point rather late in the game. It is the drift that must be constrained from the outset, indeed zeroed out to the extent possible, in order to protect the margins intrinsic to the initial space of safe operation.

Amalberti describes the initial space of safe operation as designed around rules and procedures, including various constraints and fail-safe procedures, all of which are intended to serve as defenses against human error and constrain possible human actions.[30] Standard operating procedures are designed to protect margins that can absorb variation in external influences, random deviations, and normal

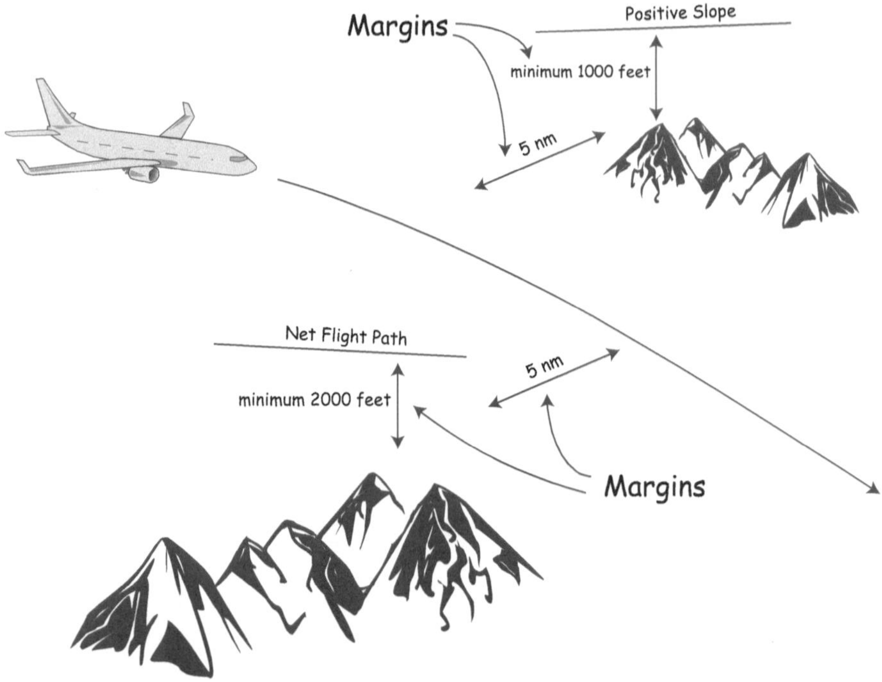

FIGURE 1.4 FAR 121.191 enroute driftdown margins.

imprecision. Well-designed procedures are constructed with carefully crafted constraints and error traps built into them; these error traps are intended to capture the inadvertent excursions into the margins, first arresting and then ejecting human error harmlessly over the side.

But the margins exist independently of procedures, and it is the margins that bound the safe operating space. We track the centerline of the runway for a reason. While nothing particularly bad will happen simply by landing with the right main wheels to the left of the centerline, something rather different may occur when the left main tires and/or brakes fail, which of course, we will have no reason to see coming, or perhaps a wind gust combines with less-than-stellar tire friction and directional control. We track the centerline to remain centered within the safe operating space and protect the seventy-five feet between where we are and the rather pronounced end of that margin … the edge of the pavement.

The margins, when they are engaged, function more or less in the dark, which is not the same thing as saying they function in a vacuum. This is a real fly in the ointment for the bean-counting population, themselves stuck in a paradigmatic mode of thought, their own version of "naive Newtonian scientism", because they cannot measure the success of safety, or the value of the investment in safety, without being able to "see" at least a representative sampling of these unfulfilled catastrophes. Pilots, too, can easily misinterpret excursions into the margins, beginning with the premise that any landing you can walk away from is a good one.

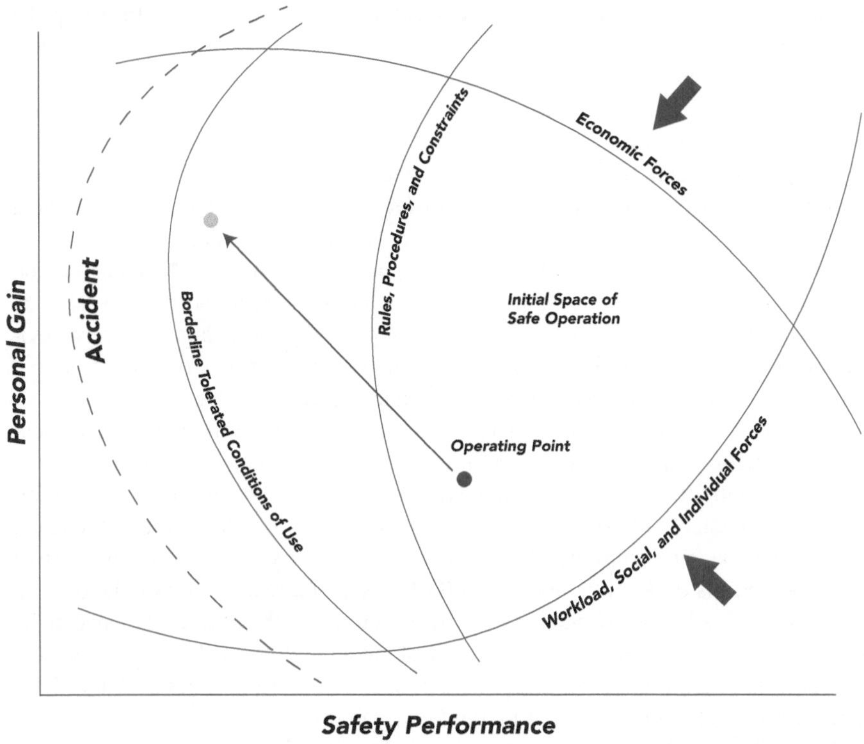

FIGURE 1.5 Amalberti's envelope of possible actions.

Procedures are, of course, eminently malleable, and thus become irresistible to management as tools for the exertion of control over employees, obfuscating the true role procedures play in the work of constructing and protecting the margins. The almost reflexive response to a mishap in an audit-centered universe is to install one more line of code, still banking on Simon's premise of tweaking "the same knowledge and information" so that "Two persons ... can rationally decide only upon the same course of action".[31]

On the other hand, a willful imposition of drift toward the borderline conditions can be used as a means of usurping top-down control and preserving an identity of craftsmanship. In conventional human factors parlance, this willful imposition of drift is made up of routine and optimizing violations. In fact, rather than demonstrating craftsmanship by artfully bending and folding procedures, the craftsmanship inherent in professional airmanship actually lies with the skill used to protect the margins and remain squarely within the safe operating space when organizational and social forces exert strong pressure to drift over somewhere near the edge, the borderline tolerated conditions of use.

Strategy is always anticipatory, always dynamic, and never assured. But for the simple reason that you cannot think of a thought until you think of it, strategy operates in a kind of bilateral reciprocity with the concept of resilient performance. Resilient performance is the anti-matter to persistent vulnerability, wild enough to trump the joker. In his work on high-reliability organizations, Karl Weick has defined resilience, at the organizational level, as the organization's "capability to investigate, learn, and act *without* knowing in advance what they will be called to act upon".[32] This definition is easily migrated from an organizational level to the individual and crew levels. Weick expands his definition to say that resilience is the "capacity to do quick study, to develop swift trust, to engage in just-in-time learning, to imagine detailed next steps, and to recombine fragments of potentially relevant past experience".[33]

Strategy and resilience are embodied in the concept of prudence. Neither function independently, and both require that we replace the misinterpretation of experience with mindfulness, as well as anticipate the threat posed by the normalization of deviance. Every flight will be different, and every flight will experience variability in human performance as well as an open environment subject to considerable modification.

When I'm having a particularly irreverent day, I've always thought that the real goal of a professional pilot is to arrive at the Pearly Gates, checking in for eternity, and find yourself standing in line just ahead of one of those bean-counters. The fellow at the desk checks off your name, and as you are about to step through the Gate, he says, "Hey, hang on a minute. We've got something for you …". He reaches into a cabinet behind his desk, pulls out a fat roll of newsprint, reaches across the desk to hand it to you, inadvertently clubbing the bean-counter in the back of the head as he does so. The roll of newsprint turns out to be about a hundred yards long, containing line after line of ten-point font, each line a description of every disaster that did not happen because of your lifetime of meticulous airmanship.

Alas, in the earthly scheme of things, a more practical axiom, one that we could use to replace the classic tenth axiom, may be that suggested by Karl Weick, when he said that reliability is a dynamic non-event.[34] We have to be careful not to confuse reliability with safety; they are not necessarily the same thing. However, insofar as safety is the non-occurrence of an unsafe event, a non-event, we get a much more meaningful axiom if we say that *safety is a dynamic non-event.*

So how does our meticulous pilot facilitate a dynamic non-event? Start by not misreading James Reason. We work with a structurally deterministic system in a highly complex environment, while remaining entrusted with an almost completely autonomous free will. A system approach does not obviate the pilot's singular responsibility as the final authority over the operation of the aircraft. As a first guideline, we can use that authority in a manner to protect the margins. We can leverage human adaptability to ensure that we remain well inside the space of safe operation. We can construct and modify a strategy aimed at ensuring that what can go right, does go right.

But we must also integrate within the strategy an awareness of mistakes that we do not want to make. From the outset, we must anchor the execution of the authority vested in the pilot-in-command by understanding when, and how, to say no.

NOTES

1. U.S. Civil Aeronautics Authority, *Civil Aeronautics Bulletin No. 5, Flight Instructor's Manual*, June 1, 1939.
2. Ibid.

3. Herbert Friedman, Ada Kera Friedman, "The Legacy of the Rockne Crash", Aeroplane Magazine, May 2001.
4. Civil Aeronautics Board, *Of the investigation of an accident involving aircraft of United States registry NC 21789, which occurred near Lovettsville, Virginia on August 31, 1940.*
5. James Doolittle, Carroll V. Glines, *I Could Never Be So Lucky Again,* Bantam Books, 1991.
6. John W. Thompson, "Studies to Develop Methods of Determining Emotional Stability", Civil Aeronautics Authority, Bureau of Federal Airways Technical Development Division, Note No. 18, Washington, D.C., 1939.
7. Jack W. Dunlap, "Test of 'The Ability to Take It'", Civil Aeronautics Administration, February 1943.
8. Thompson.
9. Ibid.
10. Alex Varney, *The Psychology of Flight,* D. Van Nostrand, New York, 1950.
11. U.S. Civil Aeronautics Administration, *C.A.A. Technical Manual 106, Pilot Instruction Manual,* July 1958, 14.
12. Herbert Simon, "The Proverbs of Administration", *Public Administration Review,* Vol. VI, No. 1, Winter 1946, 64.
13. James Reason, "Human Error – Models and Management", *British Medical Journal,* Vol. 320, March 18, 2000.
14. Ibid.
15. Karl Weick, Kathleen Sutcliffe, *Managing the Unexpected: Assuring High Performance in an Age of Complexity,* Wiley, 2015. Kindle Edition, 7.
16. Sidney Dekker, *The Field Guide to Understanding 'Human Error'.* Ashgate Publishing Ltd, 2014. Kindle Edition, 1.
17. James Reason, *Human Factors: A Personal Perspective,* Human Factors Seminar, Helsinki, February 13, 2006.
18. Sidney Dekker, "What Is Rational About Killing a Patient with an Overdose? Enlightenment, Continental Philosophy and the Role of the Human Subject in System Failure", *Ergonomics,* Vol. 54, No. 8, August 2011, Taylor & Francis, www.tandfonline.com, 681.
19. Jerome Bruner, *Actual Minds, Possible Worlds,* Harvard University Press, 1986, Chapter 2, "Two Modes of Thought".
20. Ibid.
21. Jerome Bruner, "Culture and Human Development: A New Look", *Human Development,* Vol. 33, 1990, 344–355.
22. Ibid.
23. Bruner, *Actual Minds.*
24. Bruner, "Culture and Human Development: A New Look", 344–355.
25. Stanley N Roscoe, *Aviation Psychology,* Iowa State University Press, 1980, 162–163.
26. Ibid.
27. Rene Amalberti et al., "Violations and Migrations in Health Care: A Framework for Understanding and Management", *Quality and Safety in Health Care,* December 2006, doi: 10.1136/qshc.2005.015982.
28. Thompson.
29. Amalberti.
30. Ibid.
31. Simon.
32. Weick, 94.
33. Ibid.
34. Ibid., 17.

2 On Just Saying No

We all have our mentors, people who figure prominently in our apprenticeship, people who set the bar, and to whom we look for the experience we don't yet have. One of those in my early career was a fellow by the name of John Brown, a retired Air Force C-141 and KC-135 pilot from Arkansas, in his mid-fifties, with thinning red hair, red eyebrows, and red eyelashes ... the only way you could tell whether John was awake in the red cockpit lighting was when the end of his cigarette glowed. He was the chief pilot at the small cargo airline that had hired me as a very inexperienced first officer. We were flying former Air Force T-29s, which of course John had also flown while in the Air Force, that had been converted to the civilian type certificate of the Convair 240. No pressurization, no hi blowers, but a pair of Pratt R-2800s that still had, and needed, anti-detonation water injection. We operated scheduled routes every day; in those days, UPS did not have its own airplanes and FedEx was flying personal letters around in Falcon 20s, which we thought was almost as dumb an idea as the blue potato chips handed out some years later by another upstart airline. Flying Tigers, on the other hand, competed with us, or rather we competed with them. They flew a 747.

Captain John was known for a small volume of acerbic witticisms delivered with only the slightest hint of dry humor. I will never forget such gems as "You know, Steve, we're not aimin' for those green lights down there ...". His mentorship of me started out on my initial checkride, done in the airplane, of course, with the Fed sitting on the jumpseat, literally breathing down our necks. I was an experienced light airplane instructor by then, with a grand total of about fifteen hours of multiengine time. There we were, pounding along on the inbound leg of the procedure turn during the NDB approach. I scrutinized the ADF needle and concluded it was time to turn inbound. But the yoke wouldn't move.

I looked over at John, who was hunkered down in the left seat, staring off out into space through the left side window, smoking a cigarette. The yoke still wouldn't move. A small light inside my brain suddenly flickered and then illuminated; the Convair had an aileron-rudder bungee type of interconnect, meant to mitigate adverse yaw in the days before yaw dampers. I suspected that John had his foot planted on the rudder pedal; I decided to try again in another few seconds. When I did, the yoke turned and I rolled out onto a perfect inbound course. After we landed, John pointedly said that I had done a good job; I never did find out if he meant that I flew well or that I just knew when to keep my mouth shut. Both, it turned out, were useful skills in the cargo business.

One really dismal winter morning a couple of years after that first checkride, John and I were operating a flight from Cleveland up to Buffalo, with the intention to stop at Buffalo, then continue on to Toronto, and then return to Buffalo. This was advantageous to the customer because it was quicker and easier to fly the cargo over the border than it was to truck it across the Peace Bridge. That morning, however, the weather was abysmal. We arrived in the vicinity of Buffalo to find the RVR for

DOI: 10.1201/9781003369677-2

runway 23 was one thousand, variable to two thousand in snow. The winds were gusting to 30 knots or so at about forty-five degrees to the runway.

We held for a while, and when the RVR finally lifted adequately, we flew the approach and landed in a cloud of blowing snow, enveloped in our own private white-out when we reversed the big old Ham Standards and left them there in lieu of touching the brakes. As we emerged from the cloud of snow, we turned left off the runway while our station manager called repeatedly, and worriedly, asking where we were. John barked over to me, "Tell him to just give me a minute, fer Christ's sake … I don't even know if I'm on the damn airport yet …".

The taxiway surfaces were slick and the route over to the ramp was substantially downhill, so we were in no hurry to get there. After we parked and shut down, I stepped back out of the cockpit, kicked open the door, dropped the ladder, and started down to the ramp. When my legs got below the shield of the door, the wind literally blew the back of my parka up between my knees. It was howling. We leaned into the wind as we traversed the ramp and ducked into the trailer used for operations.

With the trailer door shut, the wind quieted and we rubbed our hands over the electric heater. John simply drawled to the young station manager, "Don't even bother to open her up, Al. We ain't goin' to Toronto".

Al, a young kid of 19, recently married and with a very new son, was perfectly happy to stay inside that morning; he knew fully well that the only way you could open the cargo doors was by standing on the tines of the forklift, and once the wind got hold of one of those doors, whoever was in its way would get flicked off that forklift like a bug. John called the home office and told the flight follower, who somewhat nervously accepted the news. Arrangements would have to be made to truck the freight after all … both ways. And of course, everyone knew what would happen next.

Sure enough, about ten minutes later, the phone rang. It was Jim, the owner. I didn't have to hear the other end of the line to know what was said. I had opened this station a few months earlier and hired Al myself. Jim and I had frequent telephone conversations, in a manner of speaking, and I could pretty much guess at the gist of the conversation. After a few minutes of subdued, murmured conversation, John hung up the phone, walked over to where Al and I were standing, and said, "Well, boys, we're not goin' to Toronto today. I already made that decision … but now Jim's blessed it …". Not a man given to much humor or emotion, I did detect a bemused twinkle in his eye, perhaps a very slight bit of an eye roll, and that said it all.

Herein lies the fractious interface between the pilot-in-command and the pilot-as-employee. It can lead to some pretty polarized opinions. As a young pilot, or really a pilot of any age just trying to pay the bills, it can be sort of intimidating, to put it politely.

There are several places within the Federal Aviation Regulations (FARs) that define the authority and obligation of the pilot-in-command. FAR 91.3, shown in Figure 2.1, is the benchmark; it is here that the words "final authority as to the operation of the aircraft" are found today.[1] These words, like so many in the FARs, have been in place for decades. Beginning in the 1930s, the "first pilot" was, by regulation, in command of the aircraft during flight. In April 1945, Civil Air Regulation 60.100, Authority of Pilot, stated that "the pilot in command of an aircraft shall be

FAR § 91.3 Responsibility and authority of the pilot in command.

(a) The pilot in command of an aircraft is directly responsible for, and is the final authority as to, the operation of that aircraft.

(b) In an in-flight emergency requiring immediate action, the pilot in command may deviate from any rule of this part to the extent required to meet that emergency.

FIGURE 2.1 FAR 91.3, responsibility and authority of the pilot-in-command.

directly responsible for its safe operation".[2] A little over two years later, in August 1947, the rule had been revised to state that

> The pilot in command of the aircraft shall be directly responsible for its operation and shall have final authority as to operation of the aircraft.[3]

And so it remains today. It is not a unique idea; Annex 2 of the International Civil Aviation Organization (ICAO), Rules of the Air, states in paragraph 2.3.1 that

> The pilot-in-command of an aircraft shall, whether manipulating the controls or not, be responsible for the operation of the aircraft in accordance with the rules of the air, except that the pilot-in-command may depart from these rules in circumstances that render such departure absolutely necessary in the interests of safety.[4]

The ICAO language further codifies the authority of the pilot-in-command in paragraph 2.4:

> The pilot-in-command of an aircraft shall have final authority as to the disposition of the aircraft while in command.[5]

Language similar to that of the ICAO Annex is used by many aviation authorities around the world, or, as in the case of Canada, the policy of adherence to ICAO standards leads to the ICAO language being regulatory. Encoded here is that singular autonomy, that point of application for free will, that has not changed through eons despite all of the changes in the architecture of man-machine interface as well as the changes in management theory and even the emphasis on crew resource management. The pilot-in-command is directly responsible and has the final authority. This forms one anchor of the paradigm.

The second anchor of the paradigm, for air carriers, is the concept of operational control. The FAA defines operational control as "The exercise of authority over initiating, conducting, or terminating a flight".[6] Embedded within this concept is the obligation of the certificate holder to protect the public safety. The concept of operational control dates to 1953 and that year's revision to CAR 40.351.

> Operational Control. The air carrier shall be responsible for operational control.
>
> a. Joint responsibility of aircraft dispatcher and pilot-in-command. The aircraft dispatcher and the pilot in command shall be jointly responsible for the preflight planning, delay, and dispatch release of the flight in compliance with the applicable regulations of this subchapter and operations specifications.

b. Responsibility of dispatcher. The aircraft dispatcher shall be responsible
   1. For monitoring the progress of each flight and the issuance of instructions and information necessary for the continued safety of the flight.
   2. For the cancellation or redispatch of a flight if, in his opinion or the opinion of the pilot in command, the flight cannot operate or continue to operate safely as planned or released
c. Responsibility of pilot in command. The pilot in command shall during flight time be in command of the airplane and crew and shall be responsible for the safety of the passengers, crew members, cargo and airplane.[7]

Buried in this text is a phrase, actually a word, that is critical to understanding both the responsibility and autonomy afforded to the pilot-in-command. The word is opinion; this original construction of operational control pivots on two opinions, that of the dispatcher and of the pilot-in-command. It is impossible to know precisely what meaning the authors of this language intended; meanings, inferences, and implications can change over just a few years. But an early idea behind the word opinion, as it was imported from Old French, associated it with "judgments founded upon probabilities".

This regulatory language was promulgated not long after Herbert Simon proposed that two people, in possession of the same information, could not help but arrive at the same conclusion. Yet the word opinion does not seem to tolerate that notion; implicit within a judgment founded upon probabilities is an interpretation of evidence through the lens of experience, introducing substantial subjectivity, albeit a subjectivity that must be informed. The purpose behind the use of subjectivity is to account for facts not yet in evidence. At its most basic level, the mode of thought used most often when considering economic implications, the paradigmatic mode, cannot account for facts not yet in evidence. The use of judgment founded upon probabilities introduces the narrative mode of thought, which, interpreting evidence through experience, anticipates the story, anticipates the peripeteia, the departure from a canonical state, and thus steers the story in the correct direction.

This concept is today found in FAR 121.533, Responsibility for Operational Control, Domestic Operations, shown in Figure 2.2. This rule states that, among other things,

> The pilot-in-command and the aircraft dispatcher are jointly responsible for the preflight planning, delay, and dispatch release of a flight in compliance with this chapter and operations specifications.
>
> Cancelling or redispatching a flight if, in his opinion or the opinion of the pilot-in-command, the flight cannot operate or continue to operate safely as planned or released.
>
> Each pilot-in-command of an aircraft is, during flight time, in command of the aircraft and crew and is responsible for the safety of the passengers, crewmembers, cargo, and airplane.
>
> Each pilot-in-command has full control and authority in the operation of the aircraft, without limitation, over other crewmembers and their duties during flight time, whether or not he holds valid certificates authorizing him to perform the duties of those crewmembers.[8]

Part 121 Supplemental, Part 135, and Part (91(k) regulations specify operational control somewhat differently because there is no aircraft dispatcher; Part 121 flag rules are nearly identical to the domestic rules. Operational control, however, actually originates within the certificate holder's management structure, typically with the Director of Operations. Larger operators may establish a broader structure of management responsible for operational control. In all cases, one of three systems for operational control must be established through Operations Specifications. This codifies the certificate holder's obligation to protect the public safety.

Notice that the "opinion" of the dispatcher or the pilot-in-command has remained in the FAA regulatory language defining operational control. It has also survived in another rule, one which originated in May 1940 as CAR 61.7114, Flight Equipment. The original language said

> An air carrier shall not dispatch an aircraft in air transportation unless the equipment required by the Civil Air Regulations for the particular type of operation involved is installed in such aircraft and in serviceable condition and, if any part of such equipment becomes unserviceable in flight, a landing shall be made either at the nearest suitable landing area where a safe landing may be made or, at the next point of intended landing, whichever in the opinion of the pilot and dispatcher is the safest procedure ....[9]

Today, this language is found in FAR 121.627, Continuing Flight in Unsafe Conditions:

> No pilot in command may allow a flight to continue toward any airport to which it has been dispatched or released if, in the opinion of the pilot in command or dispatcher (domestic and flag operations only), the flight cannot be completed safely; unless, in the opinion of the pilot in command, there is no safer procedure.[10]

With a similar history, FAR 121.629 also requires that the pilot have an "opinion":

> No person may dispatch or release an aircraft, continue to operate an aircraft en route, or land an aircraft when in the opinion of the pilot in command or aircraft dispatcher (domestic and flag operations only), icing conditions are expected or met that might adversely affect the safety of the flight.[11]

Finally, in the same revision of 1953 that introduced operational control, CAR 40.359 also introduced the suspension of operations. This rule originally stated that

> When conditions exist which constitute a hazard to the conduct of safe air carrier operations, including airport and runway conditions, the air carrier shall restrict or suspend operations until such hazardous conditions are corrected.[12]

In the modern rendition, the rule appears almost unchanged as FAR 121.551:

> When a certificate holder conducting domestic or flag operations knows of conditions, including airport and runway conditions, that are a hazard to safe operations, it shall restrict or suspend operations until those conditions are corrected.[13]

Together, these rules, and their equivalents in Parts 135 and Part 91K, form a structure of authority designed to make the necessary and appropriate decisions regarding

FAR § 121.533 Responsibility for operational control: Domestic operations.

c) The aircraft dispatcher is responsible for -

> (3) Cancelling or redispatching a flight if, in his opinion or **the opinion of the pilot in command,** the flight cannot operate or continue to operate safely as planned or released.

(d) Each pilot in command of an aircraft is, during flight time, in command of the aircraft and crew and is responsible for the safety of the passengers, crewmembers, cargo, and airplane.

(e) Each pilot in command has full control and authority in the operation of the aircraft, without limitation, over other crewmembers and their duties during flight time, whether or not he holds valid certificates authorizing him to perform the duties of those crewmembers.

FIGURE 2.2 FAR 121.533, responsibility for operational control: Domestic operations.

whether a flight can operate safely. In all cases, the pilot-in-command remains the final authority. These relatively specific requirements, in turn, introduce a wide range of required knowledge regarding conditions that are a hazard to operations or what conditions would preclude a flight from being completed safely. It is incumbent on both the pilot-in-command and the dispatcher (or director of operations if a dispatcher is not required) to be completely familiar with these conditions, both prior to and during the flight.

The FAA definition of operational control is widely used by aviation authorities around the world, with occasional modification; for example, EASA expands the language to say, "Operational control means the responsibility for the initiation, continuation, termination or diversion of a flight in the interest of safety".[14] However, while the FAA is quite clear about what operational control is and where it resides, there is considerable ambiguity in many other nations' regulatory language. A key distinction lies with the joint authority and responsibility for operational control during flight. While FAR 121 requires that the dispatcher or director of operations share joint responsibility for operational control with the pilot-in-command during flight, many authorities do not specify a joint authority and responsibility during flight; rather, the responsibility for operational control falls squarely on the shoulders of the pilot-in-command, usually from the time he or she signs the flight release. Many regulatory authorities are more specific than the FAA regarding the types of information pertaining to each flight that the pilot-in-command must be familiar with, however the use of the words "if, in the opinion of the pilot-in-command" is relatively unique to the FAA's regulatory language.

That said, there is a continuity that runs throughout all of these regulatory requirements. Essentially, the pilot-in-command has an obligation to be aware of everything that may impact the safety of flight, and there is an implicit decision regarding the continuation, or conduct, of the flight that must be made with each change in conditions. All national authorities require that the operator designate a pilot-in-command for each flight and that pilot is vested with both the final responsibility for the safety of the flight and a singular authority over all other crewmembers, as well as operational control, to ensure that safety is preserved. At the end of the day, only one person releases the brakes before taxi … the pilot-in-command.

Arriving at Baltimore one night, I was scheduled to change aircraft from one Fairchild Metro to another. Upon entering the new aircraft, I discovered that the aircraft oxygen system had been placarded as inoperative. It wasn't just inoperative; there was no oxygen bottle installed. The approved MEL allowed this at the time. I refused the airplane.

Well, that created a bit of an uproar. My point was that, despite the MEL, the airplane flight manual emergency procedures required the use of oxygen in the case of smoke in the cockpit. Pressurization was irrelevant. In my opinion, it was not safe to operate the airplane in revenue service without oxygen available for the crew. I had no problem with deferring it for the passengers and staying below 10,000 feet; but not for the crew.

The next day, the company put out a short-lived memo stating that the requirement to don oxygen in the cockpit smoke emergency procedure only applied if the system had not been deferred under the MEL, a rather interesting interpretation of the Airplane Flight Manual. Nonetheless, the director of operations supported my decision, which certainly made my life easier. But, as John Brown had made clear that day in Buffalo, the captain's decision is not dependent on official support. It is, however, dependent on knowing the airplane and the operation.

A couple of months later, I did enjoy a brief moment of certitude. The union had asked me to attend the FAA Flight Operations Evaluation Board meeting in Kansas City and to present the argument that the MEL was in error. Over the objections of several air carriers, the FAA listened to my argument and summarily ruled in favor of it; the certification office engineer chairing the proceedings was somewhat embarrassed that the situation had ever existed at all.

As a 727 first officer at TWA, I watched the captain engage in a rather heated telephone conversation with the folks in maintenance when he refused an aircraft that had one of the two yaw dampers deferred. We were in St. Louis, heading for Miami and then up to New York. His point was that we would be flying up the Atlantic route from Miami, somewhat off shore until reaching Wilmington, North Carolina. There were gulf stream thunderstorms already building out there, and the MEL limited the altitude to 26,000 feet with only one yaw damper operative. The captain wasn't interested in poking around the gulf stream that day at 26,000 feet. He wanted another airplane.

In a similar situation, I arrived one night to find that my 757 had an inoperative APU. We were heading to London from New York. The company was brand new; we had no authorization for extended range operations yet and were still required to use what were known as the blue spruce routes, up over Greenland and Iceland. This ensured, theoretically, that a diversion airport was always within sixty minutes flying time in case anything happened.

But on this night, the company's idea of a route was to use Narsarsuaq, Greenland, as one of the alternates. While this was a legal, approved alternate, it was not very practical. Air Greenland was operating a 757 into Narsarsuaq at the time, but they considered it quite a challenging operation. The airport sits at the bottom of a fjord, and the surrounding terrain is pretty mountainous. The approach is difficult, requiring a pretty steep glide path, and transport aircraft arrivals generally only land on runway 7 and takeoff on the reciprocal runway 25. Because of the single engine climb requirements, the option of a single engine missed approach is off the table inside of a couple of miles from the runway threshold. (Some months after this particular evening, the chief pilot at Air Greenland pointed out that pilots operating into Narsarsuaq often had difficulty meeting the parameters of a stabilized approach).

So, while it looked good on paper, it was pretty well-understood that you didn't need to try your first approach into that airport in the middle of the night with two engines, let alone one engine inoperative. We accepted this plan on the premise that we were operating an aircraft designed and equipped for extended range operations, and the crews all had previous experience with ETOPS before coming to work at our fledgling airline. If we really needed to, we could always execute a longer diversion to Gander or Keflavik rather than attempting the approach into Narsarsuaq.

Except in this case. Part of the premise of our acceptance was that the airplane was in a condition "equivalent" to the extended range configuration that would be required once we had gained FAA approval for such operations. At the time, the MEL required the APU to be operative at the time of dispatch for ETOPS. It was not technically required for our planned operation; but if we wanted to keep that two-hour diversion back to Gander or ahead to Keflavik in our pocket, we wanted a configuration that was consistent with the requirements for ETOPs that existed at the time (they have since been changed; the APU is no longer required at dispatch for one-hundred-twenty-minute ETOPS under the 757 Master MEL …).

I refused the airplane, and this again led to a very heated telephone conversation. My decision, of course, stood.

Years later, I made a similar decision in refusing an airplane in Minneapolis when it arrived with the captain's windshield heat circuit breaker popped. I had already reviewed the maintenance history, and I knew that the breaker had popped five times in the last four days. I was open to fresh ideas, but a routine check and reset wasn't going to cut it. While the first officer and I had dinner before the airplane arrived, we discussed this item. Predictably, in the bedlam of a late arrival an hour later, with three nearly simultaneous boardings in the same boarding lounge, neither one of us checked the circuit breaker. Fully boarded, we started the Before Takeoff checklist, got to the part about windshield heat, immediately detected a cold windshield, looked at each other, rolled our eyes, and then looked at the breaker. Sure enough, it was popped. We probably could have executed that a bit more gracefully, but the decision was made … everybody got off and we found another airplane.

In the latter case, working for a very large, established airline, there was no heated telephone conversation. There was only the maintenance controller agreeing with my decision while expressing some disbelief at the history. In the former case, I was working for a brand new, trying-to-get-started airline. These contrasting situations lead to very different responses and tensions; yet, the pilot retains the final authority and must make the decision.

But that decision can be a weighty one both in terms of the obvious, the safety of passengers and crew, and the less obvious, the economic consequences. Just digging in your heels because you can is not a professional approach to these problems. Rather, the approach involves the careful practice of authority based on technical excellence and what is known, to the high reliability people, as a sensitivity to operations. We'll discuss that more in Chapter 10.

A week after I refused the 757 with the inoperative APU, I got a phone call from the dispatcher a few hours before my scheduled departure from New York. Yet another APU was on the blink; she wanted to know if I was going to take it or not. With some intelligent discussion conducted in normal tones of voice, we determined that the route was not using Narsarsuaq, but instead Iqaluit as the sixty-minute alternate. Iqaluit is, itself, no picnic for diversion. But it has a 9,000-foot runway with a standard ILS approach. It is certainly an acceptable alternate. I saw no reason not to operate without the APU.

We were trying to start a premium class airline, and as it happened, we were supposed to be carrying a teenage member of a very well-known English-speaking nation's royal family that night. The maintenance supervisor and I discussed the implications of this young lady, cute as she was clad in her blue jeans, standing out on the ramp in the cold wind at Iqaluit simply because one of the two engine-driven generators had failed, which would have been the required outcome based on the Airplane Flight Manual emergency procedures. We managed to get the APU running after all.

But, of course, we could have re-routed the flight I had refused a week earlier up past Iqaluit as well. Nobody thought of that. We should have.

NOTES

1. United States Code of Federal Regulations, Title 14, Subchapter F, Part 91.
2. United States Civil Air Regulations, Part 60, Air Traffic Rules, Amendment 60-0, July 1, 1945.
3. United States Civil Air Regulations, Part 60, Air Traffic Rules, October 8, 1947.
4. International Civil Aviation Organization, Annex 2 to the Convention on International Civil Aviation, "Rules of the Air", Tenth Edition, July 2005, 2-1.
5. Ibid., 2-1.
6. United States Code of Federal Regulation, Title 14, Subchapter A, Part 1.
7. United States Civil Air Regulations, Part 40, Scheduled Interstate Air Carrier Certification and Operation Rules, October 1, 1953.
8. United States Code of Federal Regulation, Title 14, Subchapter G, Part 121.
9. United States Civil Air Regulations, Part 40, Scheduled Airline Certification, Interstate and Inter-Territorial, Amendment 51, June 1, 1940.
10. United States Code of Federal Regulation, Title 14, Subchapter G, Part 121.
11. Ibid.
12. United States Civil Air Regulations, Part 40, October 1, 1953.
13. United States Code of Federal Regulation, Title 14, Subchapter G, Part 121.
14. European Union Aviation Safety Agency, Annexes to the Draft Commission Regulation on 'Air Operations – OPS', Annex I, Definitions for Terms Used in Annexes II–VIII.

3 Prudence

In the summer of 1901, Wilbur Wright published an article in the German publication Illustrierte Aeronautische Mitteilungen, in which he advocated for greater safety afforded by the pilot lying in a prone position instead of dangling his feet beneath the flying machine as Lilienthal had. He argued that the safety of the operator was more important to successful experimentation than any other point. Sadly, the whereabouts of the original English version of this article are unknown, which is unfortunate because Wilbur's next statement is profoundly astute, depending on how it is translated:

> Die Geschichte vergangener Versuche beweist, dass eher grossere Vorsicht als grossere Kuhnheit nothwendig ist.[1]

A simple translation of this would be that more caution and less daring is required; however, in the 1953 edition of "The Papers of Wilbur and Orville Wright", this is translated as "The history of past investigation demonstrates that greater prudence is needed rather than greater skill".[2] This is a fascinating translation.

The modern usage of prudence retains elements of caution, judgment, and wisdom, but a much more comprehensive meaning is found in the Summa Theologica, St. Thomas of Aquinas' 13th-century work on theology and philosophy. While the relationship of the Summa Theologica with professional piloting seems, to put it politely, tenuous at best, it is worth exploring Thomas' discussion of prudence, a very small part of the Summa, in order to color in Wilbur's remark. The Wright brothers' father, Milton, was after all a bishop of the Church of the United Brethren; and while we do not know what Wilbur actually wrote in his original draft, it seems likely that he would have had at least a passing familiarity with the extensive discussion of prudence as the principal virtue in St. Thomas' treatise. It is thought provoking to contemplate how this concept may have influenced Wilbur's meaning in his writing, particularly in light of the surprisingly direct and applicable ideas contained within Thomas' conception of prudence. If we accept the translation as it appears in "The Papers", it is possible that two years before he actually flew a powered airplane, Wilbur Wright established a practical and meaningful idea of the capable and competent aviator.

The early Christian theologian St. Augustine, the patron saint of, among others, brewers and printers, once said that "prudence is the knowledge of what to seek and what to avoid",[3] and he specifically described it as "the avoidance of ambushes".[4] St. Thomas himself cited Aristotle's definition of prudence as "right reason applied to action".[5] This immediately opens the question of what exactly is right reason? Centuries earlier, Cicero, the Roman philosopher and senator, had described right reason in this way when describing the foundation of natural law:

> True law is right reason conformable to nature, universal, unchangeable, eternal, whose commands urge us to duty, and whose prohibitions restrain us from evil.[6]

DOI: 10.1201/9781003369677-3

What on earth has that got to do with flying airplanes? Not so much; rather, it has a great deal to do with the execution of the duties and responsibilities placed on the pilot-in-command. Commanding an aircraft carrying passengers is a distinct and singular matter of human trust. There is no circumstance, not even far out at sea, in which human beings are less capable of fending for themselves than when aboard an airplane without the requisite skills or tools to effect a safe return to earth. There is a moral responsibility, a duty, to execute that trust and to ensure that all who have placed their lives in the hands of the pilot are indeed returned to safely to earth. The execution of that duty has to do with right reason; prudence, as right reason applied to action, lands squarely in the lap of our capable and competent aviator.

Most of us, being of the extroverted nature discovered by the Civil Aeronautics Authority researchers so many years ago, and consequently not being terribly disposed to abstract thought, do not spend a whole lot of time contemplating this trust. We frankly take it for granted that the simple act of returning ourselves safely to earth will do quite well for all of our passengers as long as they keep their hands and feet inside at all times. On the other hand, the degree of risk that we might accept on a personal level is not the same thing that might be acceptable for one or two hundred other people, who coincidentally, don't get a vote in the decisions that we make. This is the premise underlying the differences between Part 91, Part 135, and Part 121. The decisions made when flying solo in our Cessna may incur greater risk than those we make when flying a fully loaded Boeing. How do we, as pilots, maintain the correct frame of reference?

Easy. Stand in the doorway.

For many years, I made it a point to stand in the cockpit doorway as my passengers deplaned. The company wanted me to do this; of course, they also failed to provide a turnaround schedule between legs that actually gave me the time to do it, but never mind that. I did it for an entirely different reason than the public relations spin the company was aiming at. I did it because of an awful night years ago in Richmond, Virginia.

I was living in Richmond then. I was a check airman for a regional airline flying Fairchild Metroliners. I was also the chief accident investigator for our Air Line Pilots Association Master Executive Council, or local. At about 9 PM one evening, I got one of those dreaded phone calls, of which I have had several since. We had lost an airplane on takeoff at Raleigh Durham.

I went about my duties, making all of the necessary phone calls and getting my stuff together to leave immediately for Raleigh. I left my wife and started the three-hour drive at around 10 PM. Almost immediately I realized that I owed it to my friends at the Richmond airport, the airline staff, to pay a visit before I continued to Raleigh. The airplane, you see, had been bound for Richmond. So, I aimed my truck toward the airport parking lot, just a few minutes from my house.

It has been so long now that I don't recall whether I had to actually go through security or not; way back when we could bypass it with our airline identification. But the people meeting the flight were not allowed through security at that time of night. They had to wait just outside the security checkpoint. And that is where they were when I walked up to the checkpoint and passed through, heading toward the arrival gate.

And I realized, right then, that those family and friends waiting outside security did not know. They did not have any idea why the flight was late. I did. I knew that their loved ones were not coming. Ever. Not a one of them.

When I approached the podium at the gate, I recognized Linda, the agent, a charming young mother in her early thirties. I'd known her for several years. She was looking down at the podium. When she looked up at me, her eyes were red, her cheeks wet with tears. She could barely utter, "I don't know what to do. What do I do?"

About that time, another young agent arrived in her blue jeans and winter coat, having come from home as soon as she heard. The ramp supervisor, a very capable, decisive young man named Vernon, had come up from downstairs shortly after. We conferred and, well, did the things you do in such a situation. Linda knew that station management folks were on their way. We elected to wait for them to arrive. At that point, it was time for me to go; I had a long drive ahead.

I pulled Vernon aside, asked him to stay close to the two agents, and then said, "Get me out of here. I can't go back past those people". "C'mon, man", he said as he took my elbow. He led me downstairs and out the exit to the curb from there and then headed back up to the gate.

From that day forward, I have never looked at passengers and their families in the same way. I commuted to work on my airline. When I deplaned, I was usually on a mission, headed for my next flight or directly to the hotel. I walk fast. I get irritated at people who drag themselves slowly up the jetway and particularly at people who stop right in the middle of the exit to ask the agent about their bloody connections. And then I remember how absolutely beautiful those people are and how incredibly wonderful it is that they are about to walk out past security into the arms of their families and friends. And I really don't care how long they stand there, because I remember what it looks like outside the gate when they aren't coming … as far as I'm concerned, they can stand there all day.

Thus, for me, it is not hard to equate my job with Cicero's concept of a true law, which applies to all men and is eternal and unchangeable. Many of my colleagues are caught up almost daily in a regulatory paranoia brought on by the clash between the Great Society and a society predisposed to litigious behavior. But returning our passengers safely to earth is indeed a part of the natural law, a law that transcends regulation and civil claim, a law that applies to all people.

THE STRUCTURE OF PRUDENCE

St. Thomas' exploration of "right reason applied to action" incorporates the morality of natural law with a set of attributes that are essential to executing the duties of the pilot-in-command. He builds a great deal of his case for prudence around human reason, the axis along which he attempted to synthesize a philosophy from both Scripture and Aristotle. Buried within the dialectic that comprises the Summa Theologica, Thomas expands on his discussion of reason as applied to prudence with a fascinating and strikingly prescient passage. He says that

> Yet the need for reason is from a defect in the intellect, since those things in which the intellective power is in full vigor, have no need for reason, for they comprehend their truth by their simple insight, as do God and the angels.[7]

At the risk of completely freewheeling off the rails into theological discourse, compare this passage with Sidney Dekker's concept of "naive Newtonian scientism":

> ... total knowledge of the world is achievable; the world is "out there" as an object entirely separable from observers. People will know the truth if they are fully rational, once the correspondence between the picture in their mind and reality in the world is perfect.[8]

Both descriptions, centuries apart, regardless of faith, capture the human concept of perfection and the direction of progress. Both descriptions outline a benchmark for the subtle expectations that lie in the background of human endeavor and which particularly inform the idea of a capable and competent aviator who never allows his ship to crack-up. Whereas Dekker simply describes the total knowledge idea as naive, Thomas elaborates:

> On the other hand, particular matters of action, wherein prudence guides, are very far from the conditions of things intelligible, and so much the farther, as they are less certain and fixed. Hence, although in certain other intellectual virtues reason is more certain than in prudence, yet prudence above all requires that man be an apt reasoner, so that he may rightly apply universals to particulars, which latter are various and uncertain.[9]

These passages probably need a little interpretation. By the term "intellective power is in full vigor", Thomas likely meant the same thing that Dekker describes with the term "fully rational". Comprehending the truth through simple insight, in Thomas's words, is the same thing as knowing the truth through attaining a perfect correspondence between the picture in one's mind and reality, in Dekker's words. In Thomas's view, total knowledge of the world, those things in which the intellective power is in full vigor, is equivalent to the perfection associated with God and the angels. Man, on the other hand, suffers from a defect in his intellect ... he does not have total knowledge of the world ... and thus needs reason to navigate matters of action. In Dekker's description, total knowledge of the world refers to scientism, or the notion that science can ultimately access, understand, explain, and predict everything that is, and that no other knowledge beyond science exists. Either way, both St. Thomas and Sidney Dekker agree that a perfect correspondence between the picture in one's mind and reality is beyond reach, and indeed, naive.

With this in mind, Thomas claims that uncertainty exists and can best be answered with reason, "rightly applying universals to particulars". This is a powerful assertion from a 12th-century monk who spent his life developing syntheses between the Bible and the philosophical work of Aristotle. The synthesis is crystal clear if we consider that, on the one hand, he is probably leaning toward depravity and the Fall of Man as explanations for the defect in the intellect, while on the other hand, he states that "some particular matters of action ... are very far from conditions of things intelligible", an almost pragmatic acceptance of the "defect" and acknowledgement that some situations can be deeply complicated, nuanced, and contingent.

Thomas identified three principal acts associated with prudence, stated as follows in the Summa:

> The first is "to take counsel", which belongs to discovery, for counsel is an act of inquiry, as stated above. The second act is "to judge of what one has discovered", and this is an act of the speculative reason. But the practical reason, which is directed to action, goes further, and its third act is "to command", which act consists in applying to action the things counselled and judged.[10]

Taking Counsel

The first act, "to take counsel", is an expansive idea itself, but suffice to say that counsel is an act of inquiry, deliberation, and reason, because as Thomas observed, man "is unable by simple insight to comprehend with certainty the truth of things, especially in matters of action which are contingent". He expands on the idea of counsel in this way by introducing the idea of a "contingent singular", and this is important to his thinking. He differentiates frequently between matters that he deems universal and those which he deems particular; the contingent singular is another way of referring to the particular. Yet the term contingent further describes the singular as uncertain because it is dependent on other factors. This is where he identifies the vulnerability of simple insight; we cannot keep track of all of the variables, the contingent singulars, "as do God and the angels".[11]

Thomas expounds on this problem a bit later in the text, saying that the word "counsel" means many sitting together to confer. He says that in the case of contingent singulars, it is not easy for one person to take into consideration all of the factors, conditions, and circumstances that matter to the particular situation. He argues that while one person may be able to comprehend a simple, universal situation, in more complicated situations comprised of multiple variables, greater certainty is reached by several people evaluating the same variables.[12]

When one considers the origins of CRM, this is pretty interesting. Reaching back to the early mail pilots and carrying forward through the point in time when CRM came onto the scene, it was almost implicit in the concept of a capable and competent aviator that he be able to do exactly what Thomas says man cannot do, namely, comprehend with certainty the truth of things, especially in matters of action which are contingent. Beginning with the awareness raised following the Tenerife disaster in 1977 and the United Airlines fuel starvation accident at Portland, Oregon, in 1978, it became apparent that this approach had been markedly unsuccessful; the first principal of CRM became, and remains, to take counsel.

The problem, of course, is that the words "capable and competent" are closely associated with a perceived demeanor and state of mind in which one actually *is* able to comprehend with certainty the truth of things, especially in matters of action. At the 1979 NASA conference addressing Resource Management on the Flight Deck,[13] it is fair to say that much of the industry arrived still coasting on the momentum created by the 1940's fascination with "The Ability to Take It", and rather less aware of Thomas' prescient observations regarding human limitations. The initial approach to this problem was to import techniques from management

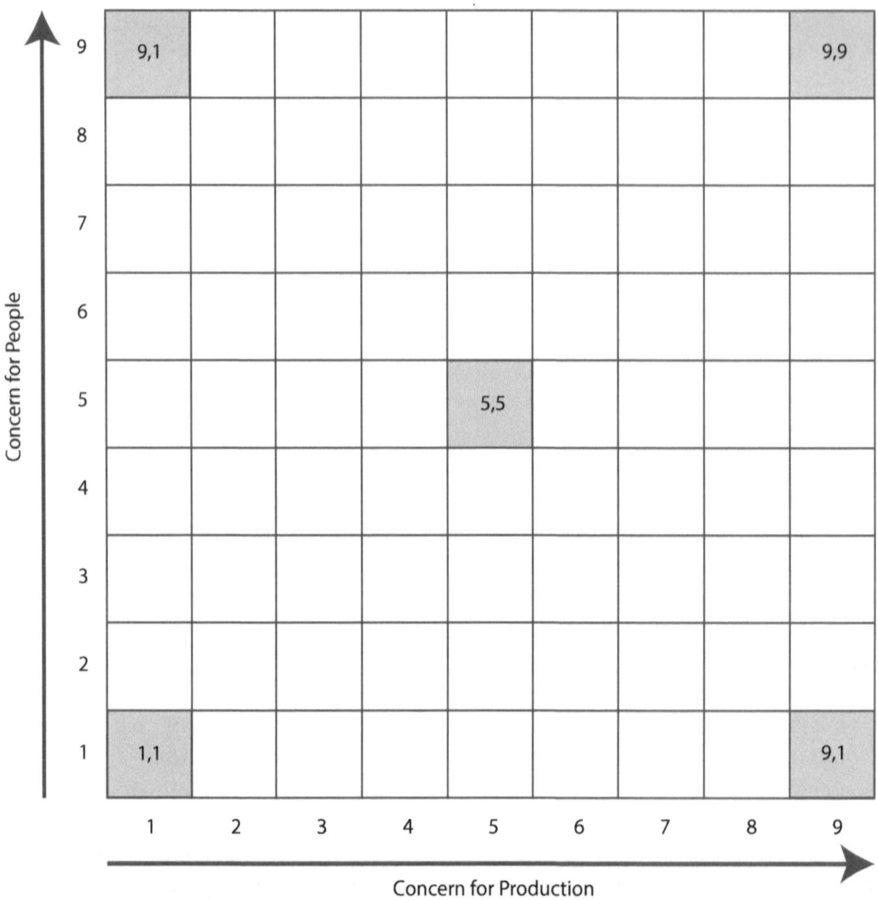

FIGURE 3.1 The Blake-Mouton grid.

schools, only then really taking off in the business world. One of these techniques was the use of the management grid developed by Blake and Mouton,[14] shown in Figure 3.1, with which most of us more wizened pilots are familiar. The grid is comprised of two axes: one represents the mission-oriented aspect of personality and the other represents the human sensitivity aspect. Predictably, our capable and competent aviator, being free of abstract thought and steeled in the ability to take it, tended to score rather high on the mission-oriented axis and not so well on the human sensitivity axis.

A certain amount of cognitive dissonance arises from the notion that one could actually learn to score highly on both axes. There was an immediate tendency, discussed at the NASA workshop, to assume that a so-called 5-5 score … half of one and half of the other … represented a pragmatic blend of personality.[15] Indeed, this was found to be typical of American corporate management, but was quickly identified at the NASA conference as a compromise that defined complacency in the cockpit.

Nonetheless, the accident record of the time was replete with aviators who tended to fall into the "9-1" category … scoring highly on mission orientation and poorly on human interaction. A notable chunk of this result may be due to the aviator-as-warrior perception that carried over from the strong military background present in civil aviation during the post-war years. Yet another aspect may be the traditions associated with language such as "final authority" and "directly responsible for the safe operation of the aircraft", language which carries over from centuries of maritime law and the authority vested in sea captains. Sea captains, after all, can perform weddings, depending on the flag of convenience, as well as override a bill of lading, refuse cargo, and incarcerate someone while aboard ship. Back in the day, they could order floggings, cut the rum ration at will, and even impress innocent bystanders into the crew, or so the stories say … not the sort of personality likely to score highly on the concern-for-people axis. Indeed, "final authority" and "directly responsible" are not terms that inhabit the middle of any scale.

CRM has come a long way since 1979, now functioning in what is essentially a sixth generation with a principal focus on threat and error management. In previous generations, CRM invested a good amount of time on communications, workload distribution, and the shortcomings of residual attention first identified in aviation psychology by Stanley Roscoe. However, the language and core concepts of management science, dating all the way back to Frederick Taylor, are primarily concerned with efficiency in production; indeed, the X-axis of the Blake-Mouton managerial grid is typically labeled "concern for production".

And the pilot who is uncomfortable with Thomas' idea of a "conference between several" is almost inevitably going to make the mistake of assuming that the variables that he is managing are universal, absolute, and simple, and that one man, by himself, is sufficient to consider these things. Herein is the problem: insofar as every flight is comprised of contingent singulars, limiting oneself to the application of universal principles will not be enough. If you have grown up believing that your intellective power is in full vigor, or if you associate that concept with your personality or identity, you will necessarily be limited to the universal principles and overlook contingent singulars. That can get you killed.

Judging of What You Have Learned

The second act Thomas describes in his discussion of prudence is "to judge of what one has discovered". He conceives of judgment as the movement toward certainty from the universal condition to the "contingent singulars" or the specific condition. "To judge of what one has discovered" specifically describes forming an opinion. Recall that Part 121 requires that the pilot-in-command have an opinion in three instances: FAR 121.533, Responsibility for Operational Control, FAR 121.627, Continued Flight in Unsafe Conditions, and FAR 121.629, Flight in Icing. In addition, FAR 121.667 says that "The pilot in command and an authorized aircraft dispatcher shall sign the release only if they both believe that the flight can be made with safety",[16] which uses the word "believe" as another way of requiring an opinion.

A well-thought-out opinion lies at the heart of all decision-making; the pilot's opinion is the cornerstone of aeronautical decision-making. Working pilots are continually judging of what they have discovered. The act of judging places a sovereignty, and an autonomy, in the pilot's hands that is inescapable. You are it; there is no one else to do it. All of the aforementioned regulations are quite clear that the pilot-in-command must judge of what he has discovered from sign in to sign out. (This is not to disrespect the work of the dispatcher; in operations conducted under applicable FAA regulations, he or she must form the same requisite opinions in the same way. The difference, of course, is that the dispatcher is planted with feet firmly on terra firma. But everything said here regarding the prudence required of the pilot-in-command is equally applicable to the dispatcher, which of course is the whole point of dual authority.)

Thomas' advocacy of prudence has to do with foresight. He quotes Isidore of Seville, stating that "A prudent man is one who sees as it were from afar, for his sight is keen, and he foresees the event of uncertainties".[17] His commentary about the deliberative aspects of the first two acts of prudence organizes them around "the end", or the intended goal … in our case, the safe return of ourselves and our passengers to earth. He then clearly builds a foundation for the application of reason, in order to properly analyze what needs to be done and in what order to achieve the desired end.

Executing Command

Once the analysis of what needs to be done has been completed, the third act comprising prudence is that of command. Going back to the original premise … that prudence is right reason applied to action … command resides within the action part of this definition. While taking counsel and reaching judgment are deliberative, applying that to action is the execution of command. In Thomas's view, command is really what people do with their free will. Command embodies the action that follows judgment.

This isn't as mystical as it sounds. I did this every day. I studied the weather, the NOTAMs, the maintenance log, read up on the deferred items and associated procedures, and then I looked at the flight plan put out by the dispatcher. This is a pretty standard package of taking counsel. Then I looked at the fuel uplift prescribed in the flight plan, and I formed an opinion, made a judgement, following which I either called the dispatcher to add fuel, or I didn't … either way is command. Indeed, I often arrived at the gate to find that the dispatcher had already exercised the three acts of prudence … taken counsel, judged of what he or she has learned, and exercised command … and there was 1,500 pounds of dispatcher add fuel already uplifted.

On the other hand, an irregular operational situation may require more counsel-taking than the standard package. These situations demand that the pilot keep open ears and listen to counsel carefully. Most importantly, the counsel of people who possess greater specific expertise with those factors pertinent to the situation is critical. The judgement rendered in these cases must be thoughtfully considered. An easy example of this may be the input of your cabin crew when dealing with an intoxicated passenger. Any flight attendant over twenty-one years old has more experience with unruly passengers than those of us in the cockpit may ever have; their counsel is critical, but it should not replace the judgement expected of the pilot-in-command.

Getting ready to fly an MD-80 from McAllen, Texas back to Dallas one day, one of the flight attendants came up to the cockpit and expressed serious concern with a strange humming noise she could hear in the aft galley area. She had never heard it before, and it bothered her quite a bit. I accompanied her back to the aft galley, and, sure enough, there was a mild audible hum. I realized almost immediately that I was in a pickle; I don't spend a lot of time in the aft galley before departure, so I have no idea what is or isn't a normal sound. My gut sense was that there was no issue. But I did not want to discount her concerns for two reasons: first, I wanted … needed, actually … her inputs for the rest of the day's flying, and I wanted her to feel like she could freely express any inputs that she had. Second, more importantly, for all I knew she was right, never mind my male-authority-figure confidence. There was no maintenance normally on hand at McAllen; we would have to call some contract guy who probably wouldn't know any more than I did what that noise was. What to do?

I lucked out. As it happened that day, several overhaul base mechanics were buried inside another MD-80 at the next gate, dealing with a major mechanical problem … you don't get the guys from Tulsa on a road trip for nothing … and I bounded across the ramp and tugged on the pant leg of some poor guy standing on a stepladder up inside the E&E bay of the busted airplane. He gladly came over and evaluated the noise, which he recognized and explained to us all. His extensive experience crawling around MD-80s at the overhaul base saved the day. Had he not been there, I was already envisioning a lengthy discussion with maintenance control, the contract mechanic, and the crew in order to resolve this issue. I really felt it was a non-issue, but I've felt that way before, and been very wrong. In general, flight attendants do not come equipped with a lack of interest in abstract thought or carefree attitudes, nor are they particularly inclined to develop an "ability to take it". They can, and often do, detect threats that might otherwise go unnoticed.

Figure 3.2 summarizes the principal acts of prudence.

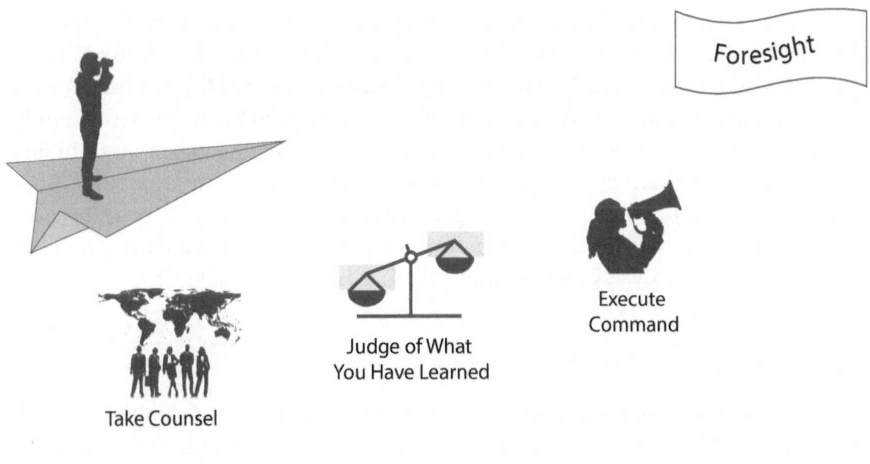

FIGURE 3.2 The principal acts of prudence.

THE ELEMENTS OF PRUDENCE

In order to expand and really dissect these three essential acts of prudence, Thomas further identified eight components of prudence, and these are fascinating when held up to the light of contemporary ideas and structures of thought. Indeed, it seems possible that Thomas' elaboration on prudence has served as the basis for much of what western civilization knows today as leadership. He placed five of these components, memory, understanding, reasoning, docility, and shrewdness, into the category of cognitive virtues. These have to do with the acquisition and analysis of knowledge. The final three, foresight, circumspection, and caution, apply to the application of knowledge, forming the basis for command.

Five of these components require closer examination; these are docility, shrewdness, foresight, circumspection, and caution.

Docility

The word docility today is associated with a meek personality, but that is not at all what Thomas intended. Of course, true history buffs will have realized by now that Thomas was not writing in modern English, but in Latin, so some of these linguistic twists have more to do with the translation of ideas than particular choices in words. By docility, Thomas was referring to a willingness to learn. This is a particularly interesting aspect of Thomas' discussion which, at least in my experience, has no specific parallel in contemporary CRM. He says that "prudence is concerned with particular matters of action",[18] and then he goes on to explain that "since such matters are of infinite variety, no one man can consider them all sufficiently; nor can this be done quickly, for it requires length of time".[19] He argues that "in matters of prudence man stands in very great need of being taught by others, especially by old folk who have acquired a sane understanding of the ends in practical matters".[20] He concludes by stating that docility indicates an openness, a readiness, to be taught, and it is therefore a crucial component of prudence. From whence, I suspect, comes the age-old axiom regarding the wisdom of our elders … and, to a certain extent, the wisdom of an apprenticeship, or even a seniority-based system of progression. However, this idea is inconsistent with much of western middle class culture. There is a strong tendency to basically stop learning at an early point in life and go on under the assumption that you already know everything you need to know. This is certainly likely to occur as experience is gained and one's expertise becomes more highly valued by peers. Regardless, at no point does Thomas suggest that docility, after some stage in one's maturation, is no longer a necessary component of prudence. No matter how elder we may be (and some of us are getting to be more elderly than we would like), we must always remain open to being taught … to learning.

Shrewdness

Shrewdness is another of these English words whose contemporary meaning can mask Thomas' intent. This idea needs to be carefully understood. In its contemporary usage, it can be confused with cunning or an intelligent approach to materialist or self-interested ends. This also is not the usage Thomas intended. Instead, he drew from Aristotle's definition: "Shrewdness is a habit thereby congruities are discovered rapidly".[21]

Thomas explained this in two ways. First, he says that

> a right estimate or opinion is acquired in two ways … first by discovering it oneself, secondly by learning it from others. Now just as docility consists in a man being well disposed to acquire a right opinion from another man, so shrewdness is an apt disposition to acquire a right estimate by oneself ….[22]

Later, he explains it another way, saying that

> a man may take good counsel, though he be long and slow in so doing, and yet this does not discount the utility of a happy conjecture in taking good counsel: indeed, it is sometimes a necessity, when, for instance, something has to be done without warning.[23]

A fine example of shrewdness under this definition lies with USAir 1549 on January 15, 2009. The aircraft struck Canadian geese when climbing out over New York after departing LaGuardia. Both engines failed.[24] The flight crew executed an extraordinary ditching in the Hudson River, resulting in no loss of life. Captain Chesley Sullenberger clearly did not have time to "be long and slow in so doing" when taking counsel; rather, something had to be done without warning, and a "happy conjecture", a congruity discovered rapidly … was that the airplane would not make it to any runway. By reaching that conclusion as quickly as they did, the crew was able to bring all of their attention and resources to bear on the problem of executing a ditching in the river.

On the other hand, there are legions of examples involving rejected takeoffs which illustrate the potential for happy conjectures to turn pear-shaped. One of the most authentic cases of this involved TWA 843, a Lockheed L-1011 departing runway 13R at John F. Kennedy in New York on July 30, 1992. The first officer was making the takeoff. Immediately after liftoff, they experienced a stick shaker, despite normal power, airspeed, and attitude; this was ultimately traced to a defective angle-of-attack vane on the right side of the airplane. However, all five occupants of the cockpit later stated emphatically that they did not believe the airplane would continue flying. Captain Bill Kinkaid elected to reduce power and land the airplane on the remaining runway, which was actually still quite plentiful. However, the configuration and speed at the time of the power reduction led to a very hard touchdown, fracturing the right aft spar and dumping tons of fuel on the rapidly heating right main brakes. The resulting fire destroyed the airplane; thanks to a superbly professional response by the cabin crew, as well as Captain Kinkead's decision to steer the aircraft off the runway to avoid the approach lights at the far end, no lives were lost.[25]

For years, the rejected takeoff demanded a very high degree of Thomas' shrewdness as the airspeed approached V1, and not every pilot on every given day may be in possession of such shrewdness. In the early 1990s, Boeing developed the RTO Training Aid,[26] which strongly advocated that shrewdness be abandoned near V1, and the decision made instead by relying on the extensive counsel regarding rejected takeoffs that the industry possessed. Statistically, there is a clear argument in favor of a decision to take the airplane into the air and resolve the failure or problem once flying safely. This counsel, rather than the previously demanded shrewdness, has been adopted industry wide in the decades since.

On March 8, 2017, the captain of Ameristar 9363, an MD-83, rejected the takeoff twelve seconds after V1, when he realized that he could not rotate the airplane. The aircraft went off the end of the runway; fortunately, there were no fatalities. The right-side elevator geared tab actuator linkage had been forced beyond its travel limits by gusty winds the night before and was locked over center. This caused the right elevator to become immovable.[27]

On October 28, 2016, the captain of American 383, a Boeing 767-300, rejected the takeoff at V1 during a takeoff roll at Chicago. An uncontained turbine disc failure had occurred in the right engine, sending a large fragment almost a thousand yards away into a UPS warehouse. The captain rejected the takeoff based on the pronounced, louder-than-might-be-expected sound of the disc failure, vibration, directional veering, *and* feeling a noticeable pressure bump simultaneously. These factors alerted him to a much more serious event than just the engine fire/failure at V1 normally practiced in the simulator. This was a very happy conjecture; there was speculation that the damage incurred to the airframe and the left engine, due to shrapnel, most likely rendered the aircraft unairworthy. It is debatable whether a successful single engine takeoff and return to landing could have been accomplished. The airplane was destroyed by fire; yet there were no fatalities.[28]

The problem, of course, is that there remain situations in which the ability of the airplane to fly is in doubt. Virtually, all company policies require a rejected takeoff in this case, but the commensurate training almost never devotes any time toward how to deploy the requisite shrewdness to make this decision. It is a difficult problem, for which qualified criteria and benchmarks would be extremely helpful; the Ameristar MD-83 was clearly not going to fly, the American 767 most probably would not have flown very far, and the TWA L-1011 would have flown quite nicely. However, as in most cases which cannot be understood or managed statistically or through other systematic means, the problem is more or less abandoned by industry. There is, therefore, plenty of room left for shrewdness as Thomas has described it.

Foresight

The word prudence derives from the Latin "providentia", which means foresight. Thomas says that foresight "implies the notion of something distant, to which that which occurs in the present has to be directed". He goes on to argue that foresight is the cornerstone of all prudence.[29]

The whole enterprise of aviation is completely predicated on Thomas's discussion of the word, because the "end" that we are directing things to is clearly a landing, and we start directing things to that end long before we get into the cockpit. The characteristic of foresight tends to be somewhat unique among pilots of jet aircraft; making a living while moving at eight miles per minute demands a very developed sense of foresight, and many people who are not so inclined may even find it difficult to understand what their neighbor, the airline pilot, is thinking during casual conversation. I remember several occasions, when coaching Little League, in which we would replace the starting pitcher after his young arm became tired or he approached his pitch limit. While the next kid was throwing his warmup pitches, I would ask my fellow coaches, "So … who do you want to go with next?" Often this was met with perplexed expressions; we had just started this kid, and he was a ways off from needing replacement. My point was that he would, eventually, need to be replaced, and my nature drives me to be planning for that. I found that most people are not disposed to think ahead like that; rather, they'll start thinking about it when the current pitcher starts throwing over the catcher's head.

The flight plan and dispatch release are documented foresight. The entire premise of flight planning is to organize contingent singulars in a manner leading to a safe return to earth. When Thomas says that "in every whole one part must be formal and predominant", he has … unwittingly, of course … defined the flight plan. The flight plan and associated documents are arranged so that, in his words, the "whole has unity".

Applying foresight takes time and practice. It cannot be done on the fly, and it does not suffer multi-tasking particularly well. For me, it took about an hour before I began the day's flying. I usually began with a review of the aircraft maintenance history (hoping that the tail number doesn't change before we get to the gate). I looked at each aircraft that we were scheduled to fly that day, because a typical multi-leg day will not afford a lot of time between legs for such contemplative study. When I detected maintenance discrepancies that had been resolved by deferral under the MEL, I looked up the item in the MEL and reviewed maintenance, operational, and dispatch actions and considerations. In this way, I was already prepared for whatever may be involved; for example, an inoperative APU means we'll need a start cart, and it also means that the ground conditioned air may be weak, and consequently I'm going to want to see the start cart in position, and have some reliable confirmation that it actually will start, before I allow the passengers to board an airplane that could be uncomfortably hot. Likewise, an inoperative fuel quantity indicator will require alternative fueling procedures (Do you know what those are? Does the local contract fuel guy know what those are?), and an inoperative air conditioning pack will limit our cruise altitude to FL 250. It turns out that one of the two pressure stop fittings on the 737 L1 door can be deferred as well, and that will limit the cruise altitude to FL 270. But that's not the same thing. With a pressure stop fitting deferred, you also have to limit the cabin differential to 6.0 psi, and to do that, you need to select a landing field elevation of 8,000 feet on the pressurization controller. This will manage the cabin differential; of course, as you descend, slowly, you will have to walk the landing field elevation down as you go.

I can tell you that, if you fail to read this fine print, the forward galley will have a squeal of about one hundred decibels. I know; we measured it. Foresight is important. The more time you have to walk through the maintenance history and required adjustments to procedure, the more likely you will *not miss anything.* (Disclaimer: unlike many of my anecdotes, this one wasn't me … I was the hapless jumpseater watching the whole thing unfold).

From the aircraft maintenance history, I would move on to review the NOTAMs. Given the amount of deteriorating infrastructure at many larger airports, this can be painstaking. I really don't care about remote taxiways or missing distance remaining signs. But there are other things.

Departing Phoenix one fine day in a 727, I was the first officer assisting a brand-new captain. Our gate was very close to the departure end of what was then runway 26R, and we always departed 26R. That day, the dispatch release was planned for runway 26L. We couldn't see any reason for this, so we requested 26R. In those days, the flight engineer computed the takeoff data, and he dutifully cranked out the hand-written takeoff performance sheet for 26R. And off we went.

It was apparent that something was amiss as soon as we rotated. Straight out in front of us, maybe a mile beyond the runway end, was a very tall, very thin crane. There was never any doubt that we would clear it … but still, the tower had to chime in with, "Hey, TWA, you gonna miss that crane?" And of course, had we lost an engine ….

I was not in the habit of reviewing all of the dispatch paperwork as the first officer in those days. This was before we had personal computers, iPads, and all that, and the captain had charge of the paperwork. The engineer and myself would get a copy of the flight plan, but not the weather and NOTAMs. The new captain had not seen anything about the crane, and I felt certain that somebody had dropped the ball on the requisite NOTAM. A week later, I was flying the same trip with a very senior and curmudgeonly captain; this time I made sure to look at the NOTAMs myself. And sure enough, there was the NOTAM for the crane. I said, "Well, I see they put out a NOTAM about that crane. I wonder when they did that?" To which the somewhat more seasoned captain replied, "Bout a month ago …"

Once the NOTAMs are reviewed, I would move on to the day's weather. I was interested in several things. First and foremost is the big picture of the weather system influencing the route or routes of flight. I used several sources to develop this characterization, such as surface analysis charts, upper air charts, the convective summary produced by the Storm Prediction Center in Norman, Oklahoma, various satellite photos and, of course, radar displays. From there, I'd move to the specific destinations we were planned for that day. I looked at the skew-T diagrams from the latest soundings and then went to the single most important source of weather information: the forecaster's discussion from the applicable weather office. Once I had digested his/her written discussion (and that may take some time), then I'd look at the TAFs and METARs. Usually the last thing I would look at is the forecast turbulence, which, by this stage, I had somewhat anticipated based on the big picture that I had memorized.

This deep dive into the day's issues … deferred items, NOTAMs, weather … leads to comprehensive foresight, from the need for a start cart through the need for an early cabin preparation for landing because of low-altitude turbulence. One of the most redeeming characteristics of this approach is that when I arrived at the gate and pulled up the flight plan, it almost invariably said exactly what I expected it to say, given my understanding of the day's issues. Of course, it did; the dispatcher had done exactly the same homework and, sure enough, reached the same or very similar conclusions. And that is how you and the dispatcher exercise the joint responsibility for operational control … by independently taking counsel, judging of what you have learned, and then converging on a plan of which to command.

Circumspection

Thomas is a bit briefer in his discussion of circumspection. He says that

> Just as it belongs to foresight to look on that which is by its nature suitable to an end, so it belongs to circumspection to consider whether it be suitable to the end in view of the circumstances.[30]

This is really speaking to situational awareness. We have to be prepared to change plans. We have to be prepared to abandon what seemed like a good plan, until unforeseen circumstances appeared. Foresight may have inspired us to add fuel for a particular alternate; circumspection is in play when we discover, while enroute, that the alternate weather has deteriorated, and we need to adjust the flight plan to specify a different alternate.

Early in the brief history of Eos Airlines, Bill Downey and I were returning from London to New York. We were flying a 757; the new company was trying to be innovative, so we had not signed up with ARINC for the typical ACARS installation. Instead, we had contracted with a new company trying to promote a newer in-cockpit system of data exchange and communication. When it worked, it was brilliant. Unfortunately, it only worked about 50% of the time. They were having a terrible time trying to get the requisite datalinks across ground stations to function consistently.

So, as we came off the tracks into Canadian airspace, we got a message from the dispatcher: need to watch storms building west of Newark. I started trying to pull up METARs for various stations west of Newark, so that I could get some idea what was where. Nothing happened. The system had failed again.

This wasn't a huge problem … yet. I thought, well, I'll get New York Radio to set up a phone patch. When I tuned in 129.9, it turned out that Air India was already on a phone patch with *his* dispatcher, and they were having a very enlightening conversation about the New York weather. So, I got what I needed by eavesdropping.

By the time we got to Calverton, a couple of hours later, New York was a mess. We were put into a hold, and we certainly did not have enough fuel to hold for very long. We had been capturing ATIS broadcasts as we came down across New England, so we had a couple of good outs; Boston, Bradley, Providence. But, without our company communications, we were limited to relaying messages to the dispatcher through the New York station by company radio.

The controller told us that some airplanes were getting into JFK via the Kingston arrival, so we started off that way. We couldn't do the whole arrival with the remaining fuel, but the routing from Calverton up north would take us very near Bradley, from which we would make a decision. So off we went.

Then we got a call from the New York station. The dispatcher wanted us to go to Stewart. Of course! I had completely forgotten about Stewart! So I told Bill, "Tell 'em we want to go to Stewart". He did, and ATC gave us a vector.

Pretty soon, Bill said, "You got any charts for Stewart?" Well, of course we do. Everybody has charts for Stewart. I looked in my Jepp binder. No charts for Stewart. When they ordered the coverage, they failed to include Stewart. You've got to be kidding.

By this time, we're down to 7,600 lbs of fuel, a bit over an hour's worth. There was weather all over to the west. It was getting dark. I looked out across the nose, and there was Albany, wide open. I said, "Time to quit screwing around … let's put it on the ground at Albany".

By the book, we needed to coordinate this with the dispatcher. But as direct communication was not possible and fuel was starting to push the limit, we exercised our authority under 121.533 to "suspend" the operation. Albany had a 7,000-foot runway, which, even in consideration of the surrounding terrain, was no different than LaGuardia or Washington National. As it turned out, this was one of my more fortuitous decisions. We were operating a premium-class airline; our 757 had 48 seats populated by very high-end travelers. While all of the big boys were up at Bradley or Boston taking a number for fuel, we were airborne from Albany within fifty minutes and on the gate at JFK forty minutes later. It almost looked like we knew what we were doing. Although there was the matter of the missing Stewart charts …

Caution

In what is probably his most interesting statement, Thomas expounds upon the idea of caution. Circling back to his earlier assertion that "some particular matters of action … are very far from conditions of things intelligible", Thomas introduces a 12th-century definition of risk management, one which presciently parallels the template we use today. He says that

> Of the evils [threats] which man has to avoid, some are of frequent occurrence; the like can be grasped by reason, and against them caution is directed, either that they may be avoided altogether, or that they may do less harm. Others there are that occur rarely and by chance, and these, since they are infinite in number, cannot be grasped by reason, nor is man able to take precautions against them, although by exercising prudence he is able to prepare against all of the surprises of chance, so as to suffer less harm thereby.[31]

This is a real foot stomper. It opens a discussion of specific risk and probabilistic risk, which form the bedrock of risk management today. That is the next chapter.

If we step back and consider the structure of civil aviation operational control, as we discussed in Chapter 2, what we find are examples of nearly everything Thomas discussed a thousand years ago. The requirement of coordination between captain and dispatcher is really a requirement to take counsel. The opinions required by FAR Part 121 and other rules are really requirements to judge of what one has learned. The authority vested in the pilot-in-command by FAR 91.3 and other rules is a codification of the command that is the necessary final part of applying right reason to action. The flight plan and dispatch release are documented foresight, and the requirement that the operator be able to contact a flight within a few minutes at any time the flight is airborne exists precisely to support circumspection. It falls to us, through competent CRM, to recognize that prudence is adaptive; you are operating in a continuous cycle of taking counsel, judging of what you have discovered, and executing command. This is a feedback loop; by incorporating docility and circumspection,

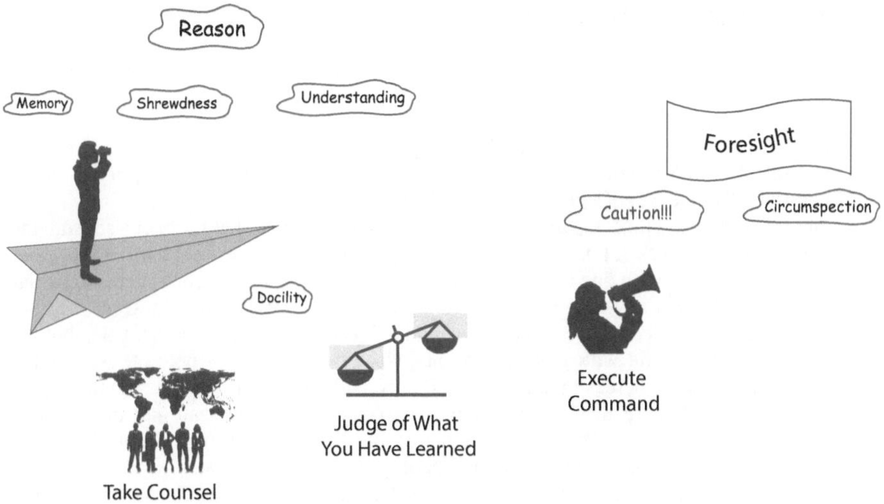

FIGURE 3.3 The elements of prudence.

the updated counsel taken, combined with updated judgements, leads to updated command. In this manner, prudence facilitates Karl Weick's dynamic non-event.

Indeed, parts of it bear an uncanny resemblance to Weick's concept of resilience; the "capacity to do quick study, to develop swift trust, to engage in just-in-time learning" … in Thomas' words, shrewdness, and docility … "to imagine detailed next steps", … foresight … "and to recombine fragments of potentially relevant past experience" … memory.

Figure 3.3 shows the elements of prudence, and how they interact with the principal acts is shown in Figure 3.2. I still find it remarkable to consider how prescient the writings of St. Thomas of Aquinas were, and still are, with respect to safety management and the execution of the duties of a pilot-in-command.

And I will always wonder which word Wilbur Wright used in his original draft.

NOTES

1. Wilbur Wright, "Die wagerechte Lage während des Gleitfluges", *Illustrierte Aeronautische Mitteilungen*, Vol. 5, July 1901, 108–109.
2. Marvin McFarland, Editor, *The Papers of Wilbur and Orville Wright*, McGraw-Hill Book Company, 1953, 61.
3. Thomas Aquinas, *Summa Theologica*, translated by Fathers of the English Dominican Province, e-artnow Kindle Edition, 2013, IIa-IIae, q.47, a1, s.c.1.
4. Ibid., q.47, a.8, ad.obj.1.
5. Ibid., q.47, a.2, ad.obj.3.
6. Marcus Tullius Cicero, *The Collected Works of Cicero: The Complete Works Pergamon Media (Highlights of World Literature)*, Pergamon Media. Kindle Edition, 2015.
7. Thomas Aquinas, q.49, a.5, ad.obj.2.
8. Sidney Dekker, "What Is Rational About Killing a Patient with an Overdose? Enlightenment, Continental Philosophy and the Role of the Human Subject in System Failure", *Ergonomics*, Vol. 54, No. 8, August 2011, Taylor and Francis, www.tandfonline.com, 681.

9. Thomas Aquinas, q.49, a.5, ad.obj.2.
10. Ibid., q.47, a.8, ad.obj.3.
11. Thomas writes, "Now there is much uncertainty in things that have to be done; because actions are concerned with contingent singulars, which by reason of their vicissitude, are uncertain. Now in things doubtful and uncertain the reason does not pronounce judgment, without previous inquiry: wherefore the reason must of necessity institute an inquiry before deciding on the objects of choice; and this inquiry is called counsel." *Summa Theologica*, q.14, a.1, ad.obj.3.
12. Here, Thomas says that "Counsel properly implies a conference held between several; the very word denotes this, for it means a sitting together from the fact that many sit together in order to confer with one another. Now we must take note that in contingent particular cases, in order that anything be known for certain, it is necessary to take several conditions or circumstances into consideration, which it is not easy for one to consider, but are considered by several with greater certainty, since what one takes note of, escapes the notice of another; whereas in necessary and universal things, our view is brought to bear on matters much more absolute and simple, so that one man by himself may be sufficient to consider these things. Wherefore the inquiry of counsel is concerned, properly speaking, with contingent singulars." Summa Theologica, q.14, a.3, ad.obj.4.
13. George E. Cooper, Maurice D. White, John K. Lauber, Editors, "Resource Management on the Flight Deck", Proceedings of a NASA/Industry Workshop Held at San Francisco, California, June 26–28, 1979.
14. Nasimul Islam, Piyali Bhattachar, "A Review on Managerial Grid of Leadership and Its Impact on Employees and Organization", *International Journal of Research*, Vol. 6, No. 10, September 2019, https://journals.pen2print.org/index.php/ijr/.
15. Cooper., 136–140.
16. United States Code of Federal Regulation, Title 14, Subchapter G, Part 121.
17. Thomas Aquinas, q.47, a.1.
18. Ibid., q.49, a.3, ad.obj.3.
19. Ibid.
20. Ibid.
21. Ibid., q.49, a.4, ad.obj.3.
22. Ibid., q.49, a.4, resp.
23. Ibid., q.49, a.4, ad.obj.2.
24. National Transportation Safety Board, AAR-10/03, "Loss of Thrust in Both Engines After Encountering a Flock of Birds and Subsequent Ditching on the Hudson River US Airways Flight 1549 Airbus A320-214, N106US Weehawken, New Jersey January 15, 2009".
25. National Transportation Safety Board, AAR-93/04, "Aborted Takeoff Shortly After Liftoff, Trans World Airlines Flight 843, Lockheed L-1011, N11002, John F. Kennedy International Airport, Jamaica, New York, July 30, 1992".
26. Federal Aviation Administration, Advisory Circular 120-62, "Takeoff Safety Training Aid", September 12, 1994.
27. National Transportation Safety Board, AAR-19/01, "Runway Overrun During Rejected Takeoff Ameristar Air Cargo, Inc., dba Ameristar Charters, flight 9363 Boeing MD-83, N786TW Ypsilanti, Michigan March 8, 2017".
28. National Transportation Safety Board, AAR-18 "/01, Uncontained Engine Failure and Subsequent Fire American Airlines Flight 383 Boeing 767-323, N345AN Chicago, Illinois October 28, 2016".
29. Thomas specifically says that "Whenever many things are requisite for a unity, one of them must be the principal to which all the others are subordinate. Hence in every whole, one part must be formal and predominant, whence the whole has unity. Accordingly, foresight is the principal of all the parts of prudence, since whatever else is required for prudence, is necessary precisely that some particular thing may be rightly directed to its end. Hence it is that the very name of prudence is taken from foresight [providentia]." *Summa Theologica*, q.49, a.6, ad.obj.1.
30. Thomas Aquinas, q.49, a.7, ad.obj.3.
31. Ibid., q.49, a.8, ad.obj.3.

4 The Nature of Risk

When I was a kid, about ten years old, my buddy and I were accompanied on one of our routine combat patrols by an eight-year-old neighbor. We went to some lengths trying to explain to our younger colleague that he would not, in fact, be able to successfully avoid being struck by a sniper's bullet simply by hitting the dirt upon hearing the rifle report. We had only recently discovered this, and we were pretty impressed by how much sense it made once we thought it through. It was really quite astounding, almost a wrinkle in the fabric of the space-time continuum conveyed through narrative causality. Something could happen that was completely unforeseen and against which all conventional defenses, in the form of reactive action, were completely useless. We were probably on the cusp of abstract thought at that age; sadly, our younger friend was not, and he was quite put off with us for disallowing his strategy of hitting the dirt during our war games. It probably didn't help that he had already been tagged as the lone enemy soldier in our three-man maneuvers.

Over the past twenty years or so, Nassim Nicholas Taleb has developed the argument that the prediction of the future, what he calls the forward process, is constrained to no more than a couple of iterations beyond the present, for two reasons. First, each predictive iteration accumulates error. The inputs to the process cannot be measured with an accuracy sufficient to prevent the introduction of tiny errors, and these eventually compound, rendering a prediction several iterations down the road hopelessly off course. Second, and more importantly, many forward processes are not linear, may not be reducible to their constituent components, and are not well described by idealized bell curve statistics.[1] This is a fly in the ointment to the narrative mode of human thought, because the prediction of a narrative relies on inductive thinking and induction cannot manage nonlinear processes. Nevertheless, largely because people are considerably more confident in their knowledge than they are actually knowledgeable, what Taleb calls epistemically arrogant,[2] an expectation of certainty is intrinsic to the forward process. The misalignment between the kind of certainty that is expected, what kind is actually possible, and what kind is ultimately fabricated, opens up the possibility of a black swan.

In the 17th century, a notion existed in Europe that all swans are white; the subsequent discovery of black swans in Australia falsified that idea, leading to a metaphoric usage of the term black swan to describe ideas that seem certain but which turn out to be incorrect. Taleb leverages this notion in describing the black swan event, which he characterizes with three qualities. First, a black swan event is an unexpected outlier; nothing in the past can clearly point to its possibility.[3] Second, it has a significant, more or less catastrophic, impact. Third, post priori, human beings immediately attempt to explain its causality.

Events which might be characterized as black swans do not necessarily appear as a black swan to all observers. In some cases, people may accurately predict an event in the same manner as a broken clock is right twice a day; but in other, more

DOI: 10.1201/9781003369677-4

meaningful cases, perspective can matter. Taleb cites the tried and proven example of the turkey. A turkey that has been consistently fed and sheltered for one hundred days has no reason whatsoever to expect what happens on the one hundred and first day, coincidentally Thanksgiving; the butcher, on the other hand, has every reason to expect what happens.[4]

Although we are generally more capable of abstract thinking as adults, the idea of a black swan remains foreign to aviation training. We built the protocols and standards for conventional flight training long before we went to the moon, before we understood system safety analysis, before we understood Reason's cheese, and most certainly before we understood Taleb's black swan. We built the protocols and standards largely based on the vast pool of "what ifs" that the veterans of the Eighth, Ninth, Twelfth, Fifteenth, and Twentieth Air Forces, among others, brought home with them after World War II. This pool of experience was necessarily biased toward those who had survived, since they were the ones who could still impart their experiences. The experience of those who did not survive was out of reach; Taleb refers to this as the problem of silent evidence.[5] The information contained within a body of silent evidence may or may not skew the conclusions drawn, but this will never be known, and the uncertainty posed by silent evidence is easily ignored in pursuit of comfortable predictions. Nevertheless, the "what ifs" following World War II mostly took the form of a "there I was" story, preserved in amber, so-to-speak, as a narrative that could be retold precisely many times over. We understood engine failures, cabin fires, navigational errors, thunderstorm encounters, and runaway propellers in this way.

SPECIFIC RISK

What-ifs form the taxonomy of a subset of risk known as specific risk. For the past several years, both the FAA and EASA have wrestled with a formal definition. To date, their proposed definition of specific risk is "The risk on a given flight due to a particular condition".[6] An even more precise definition may come from work done by Sam Pullen and his colleagues at Stanford University when writing about space-based and ground-based satellite navigation augmentation systems:

> Specific risk is the probability of unsafe conditions subject to the assumption that all credible unknown events, that could be known, occur with a probability of one (on an individual basis).[7]

In other words, what is the probability of an unsafe condition if a "credible unknown event that could be known" … an engine failure, a cracked windshield, or a failure to transfer fuel … is assumed to have actually occurred? This is the first, and most basic, concept of risk.

It is important to understand the genesis of specific risk. This is a second, more primal concept, that of risk aversion. As you read the preceding list of what-ifs, you may become aware of disappointing, uncomfortable, or even unpleasant feelings; none of us want to experience any of these events. The feelings you have are born of an aversive drive to avoid these types of events. They arise from an unpleasant or dangerous narrative, in which Bruner's peripeteia emerges from the what-if story as a specific risk for future consideration. Each specific risk that we consider represents a threat from which we have some degree of aversion; in the worst cases, of course, this means that we are unwilling to tolerate the event taking place at all. Risk aversion leads to our ubiquitous focus on the role of free will in managing threats, because aversion demands feedback-driven control. Aversion also demands a lot of Stanley Roscoe's residual attention, because it is contingent, one way or another, on being able to see the one coming that gets you, which is the only way you can assign a probability of one to any particular risk.

Another term that describes this condition is deterministic risk, in which the probability of an event is treated as finite, the inputs are known, and the outcome is observable. A process is deterministic if it can be predicted deductively based on a set of inputs and if the result is repeatable assuming the same inputs. If the probability of a particular risk is considered to be one, then one or more scenarios incorporating that risk can be modeled and specific risk mitigations can be constructed and justified. More importantly, there are clear points at which free will can be applied as a means of risk mitigation. Don't overbank when turning onto the final approach, for example, in order to avoid the possibility of an accelerated stall.

We augmented this pool of "what ifs" with an ever-expanding library of accident reports, their narrative similarities consigning most of them to the taxonomies developed from the original "what-ifs". The problem that you may have noticed is that, in order to assign a probability of one to a particular risk, indeed in order to contemplate some degree of risk aversion to any given threat, one *has to see it coming.* One has to be able to imagine it. One has to know the story, the narrative, all the way through the peripeteia to the coda, or moral of the story. This leads to the instinctive belief that accidents are predictable, repetitive, and, with the proper deployment of free will, entirely avoidable. But a risk that has been assigned the probability of one is about as far across the pond from a black swan as you can get. Alas, within the confines of a defined taxonomy, most what-ifs are probably more lions and less swans of any color.

The zebra who fails to catch the scent of the lion, fails to see the twitch of a hidden ear, or the shift in the sway of the grass, is the zebra most likely to be killed. Although it is easy to speak flippantly about such chance with catchy phrases like "You never know, today could be your last" or the ever-popular bumper sticker, "Shit Happens", we are really only comfortable with such ambivalence in cases to which we do not ascribe free will. Where free will is an operative factor, or at least where it is believed to be, chance is barely tolerated.

Our classical model of the capable and competent aviator says that he must look harder at the grass, so-to-speak, be more aware of the scent with a shift in the wind, and maintain a generally agitated level of alertness. Most zebras would agree. They have a serious aversion to lion attacks, and the lion hunts in small numbers. Lions are a countable threat; they can be detected, even if they are often not, and they can be run from, or, when the chips are really down, kicked ass-over-teakettle by a skilled zebra. In other words, the threat can be identified in real time and it is possible to counter it through specific, directed action. This represents the core of flight training, using a specific risk approach centered on "all credible unknown events, that could be known", in other words, what-ifs.

But a deterministic analysis of the lion's threat is problematic because the lion represents willful unpredictability. The lioness brings to bear a behavior, whether it be through her own free will or some other form of randomization, that introduces an uncertainty not easily contemplated in the man-machine-environment model. There is no real way to deterministically predict the probability of an attack or the probability that the attack will be successful on any given day. Yet a lion does not represent a black swan to any zebra over a week old. A wise old zebra would probably point out that the lion, as a species, is flourishing, ergo efforts to avoid being killed by lions have not been completely successful. He might go on to say that if you want to forage just about anywhere on the savannah, you will have to accept some risk. But he would also suggest that relying solely on sight, sound, and smell to detect the lioness before she springs, and the free will necessary to avoid her, is a woefully inadequate strategy. He would then point out that he has stripes for a reason.

The zebra's stripes have been studied since Darwin as an evolutionary conundrum; they must afford some benefit in the process of natural selection, but what benefit exactly? That remains to be determined with certainty. The current thinking about the stripes on a zebra facilitates the comparison of specific risk with yet a third fundamental concept, that of average risk … bearing in mind that while differentiating between these ideas is well understood within the design and certification community, they are generally not a part of the typical safety training experienced by a pilot.

The classic zebra stripe argument is that they represent a form of camouflage created by an optical illusion. Individual zebras are supposedly more difficult to distinguish in the herd, and direction of movement is also thought to be more difficult for a predator to discern.[8] No one has yet interviewed an experienced lioness on this question, but it makes sense.

The other prominent idea about the zebra's stripes is that they confuse horseflies. Tabanid flies … blood-sucking flies … are strongly attracted by carbon dioxide emission and the odor of ammonia from decaying urine. However, they are apparently put off by the light polarization encountered when looking at black and white stripes, or other patterns of spots or lines, through a mosaic eye. They have great difficulty effecting a successful landing on the zebra. This is a recent idea, developed mainly from statistical data demonstrating the overlapping of the horsefly population in Africa with that of striped animals (zebras are not the only ones), and it has been substantiated with some limited experimentation.[9] As with the lion, no one has talked with any horseflies to see if they actually give a damn about the stripes.

But the stripes-as-a-defense theory, against both lions and horseflies, creates a conveniently bifurcated idea to contrast with the more classical application of free will, looking harder at the grass and reliably detecting a lion's scent, as means of mitigating risk. The stripes represent a defense that is not contingent upon the application of any agency or free will. At the outset, it is easy to envision the stripes as an evolutionary defense against the specific risk of lion attacks; we can comprehend that narrative and we are instinctively aware of an aversion to such attacks, so it makes sense that natural selection would operate in parallel with our sense of a particular risk aversion.

The horsefly problem disrupts that paradigm, suggesting that the stripes have nothing at all to do with specific risk, but rather are evolved in a manner aimed squarely at average risk.

Certain individual horseflies still represent a specific risk; they can carry disease, which can be fatal to the zebra. If you knew exactly which horsefly to swat, the problem would be solved; they are, after all, not particularly hard to kill … as opposed to lions. But horseflies exist in countless numbers. More importantly, dozens of horsefly bites per day can literally extract a critical amount of blood from the zebra's body. The only specific, directed action for the zebra to take in defense, swatting the tail, is useful against a paltry handful of flies at a time. There is no cognitive or sensory feedback loop, so there is no control, which contrasts with the notion of looking harder at the grass and being more vigilant with respect to odors that might suggest a lion, behaviors which facilitate a sensory feedback loop leading to action. This forces a broader consideration of the risk problem and different strategies for managing it.

AVERAGE RISK

Average risk is more precisely known as probabilistic risk, and it works within a stochastic system. The word stochastic is defined by Oxford as "having a random probability distribution or pattern that may be analyzed statistically but may not be predicted precisely". A stochastic process is the outcome of a set of random variables evolving over time.

The United Nations Office for Disaster Risk Reduction explains probabilistic risk thus:

> Probabilistic risk is the chance of something adverse occurring. This method assesses the likelihood of an event(s) and it contains the idea of uncertainty because it incorporates the concept of randomness.[10]

Pullen offers this somewhat more targeted definition:

> Average risk is the probability of unsafe conditions based upon the averaged estimated probabilities of all unknown events.[11]

The concept of probabilistic or average risk is difficult to grasp for a pretty simple reason. Recall that Bruner has argued that there are two modes of human thought. The narrative mode of thought requires a peripeteia, a departure from the canonical state of things. The peripeteia is a specific risk, a credible unknown event that could be known, and thus assigned a probability of one. An average risk does not identify a peripeteia because it is not based on a single story but rather, dozens, hundreds, or even thousands of stories. There really is no story, in the traditional sense, associated with average risk. It can only be thought of in the paradigmatic mode of thought, utilizing probability distributions to forecast random behavior. We do not generally construct a worldview using probability distributions, unless you are trying to sneak out of town under crew scheduling's nose while on reserve; we prefer to do so using the narrative mode and specific peripeteias. This is almost entirely how we understand history, understand what has happened, understand our experience, and understand the what-ifs. Yet when surrounded by horseflies all day, every day, the narrative mode of thought is limited; the paradigmatic mode of thought is the necessary recourse.

The probability of any particular zebra being bitten by a horsefly on any given day is rather greater, by several orders of magnitude, than the probability of the same zebra being successfully attacked by a lioness. However, the probability of any particular horsefly bite being *fatal* to the zebra on any given day is rather less, by similar orders of magnitude, than the probability of any given lion attack being fatal. Conversely, the probability of an average daily series of horsefly bites drawing enough blood to weaken the animal is much higher. Thus, a cloud of horseflies steers the discussion toward a stochastic system, in this case the statistical likelihood that the odor of decaying urine from any particular zebra will attract a greater or lesser number of horseflies, and perhaps not just any horsefly, but one carrying disease, and that the disease will be incubated, successfully introduced into the zebra's circulatory system, and that the zebra's immune system will fail to kill the diseased organisms. Within the cloud of horseflies, we have no way of deterministically knowing when all of these conditions will be met. This is the world of probabilistic risk.

Probabilistic risk lies at the heart of Reason's world-famous Swiss cheese model. Instead of looking at a single slice of cheese and assuming there will be a large hole in the upper left corner … specific risk … we are looking at a series of slices, and estimating the probabilities that the distribution of holes across the range of cheese slices will align so as to allow an event trajectory to pass straight through. The open question is: how do you estimate the probabilities?

To tackle this, two concepts are important: the first is variability. This refers to real differences over time and space of the elements in a system and is a property of that system. How big is the slice of Swiss cheese, how big could a hole be, how many holes could there be, how close to the edge of the slice can a hole be, how long will the hole exist, how fast will it grow and shrink, and so forth. Variability can be random in nature, or it may be predictable in theory but not in a practical sense because the relationships, equations, or available data are insufficient to yield a deterministic result. The important point is that variability can be bounded by limits to the system, or it may be unbounded, in which case there are no real limits to how far an element of the system may vary. For example, a binary, yes/no type of system has only two possible variances; it is extremely limited. On the other hand, as demonstrated by Bill Gates and Warren Buffet, an individual's net worth is quite open ended. There is no systemic limit on the number of possible variances in net worth.

The second idea is uncertainty. This is the core of probabilistic risk; it refers to a lack of knowledge of the true value of some element in the system or of the relationships between several elements of the system. A lack of knowledge of the true value of some element means that a probability of one cannot be known; it can only be assigned arbitrarily in order to analyze the risk within a framework of specific risk. More importantly, in no case can a probability of zero be assigned to any value. Both the probability of one and the probability of zero define certainty.

Random variables are typically characterized, for the purposes of calculation, by a Gaussian probability distribution, often referred to as a normal distribution. This is represented in our minds as a bell-shaped curve, with the average, or mean, at the peak of the bell, as shown in Figure 4.1. The Gaussian distribution is based on what is known as the Central Limit Theorem, which essentially says that the average of a sample of the data will increasingly approximate the average of the entire

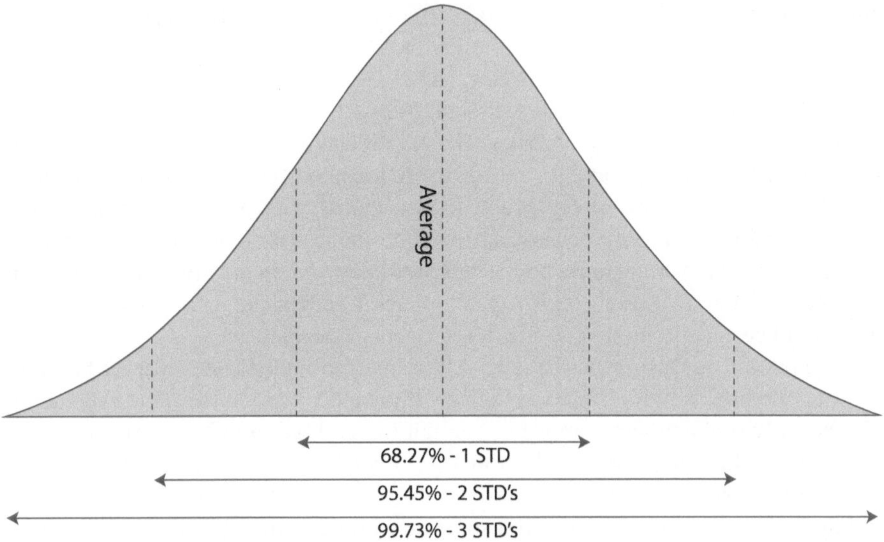

FIGURE 4.1 Typical Gaussian distribution of probability.

population of data. The more data you have in your sample, the closer the average of that sample will approach the average of all possible data … hence the bell-shaped curve. The idea is that with enough data, the mean of the data is all but immovable. This, in theory, allows statistical predictions to be made based on the probabilities of deviations from the mean. This will be familiar to anyone who has studied political polling and wondered how the opinions of three hundred people can reflect the will of the entire electorate. (Quite often, they reflect the will of the questioner, and are biased by the construction of the questions … but that is another book).

In fact, probability distributions can be weighted with historical or measured data, so may not always assume the shape of a single, symmetric bell. For example, the probability of someone's birthday being on any given date can be taken as one divided by three hundred sixty-five, or it can be calculated using historical data of actual birthdays across an entire population, which is not necessarily a symmetrical bell shape. It turns out that more people are born in the summer than in winter, for some reason. Yet the question of birthday probability is a bounded condition; there are only three hundred sixty-five possible birthdays, and it clearly falls under the Gaussian distribution purview. However, in many cases, the data used to apply a weighted correction has a strong element of variability itself, depending on how it is measured, how it is expressed or how it is interpreted. The challenge is in identifying where uncertainty exists and in determining whether the system is truly bounded or not.

I was first introduced to the concept of average risk during the work I was involved with following the TWA 800 disaster. The Fuel Tank Harmonization Working Group was an advisory committee set up in the months following the accident. We had developed a broader understanding of the conditions within an aircraft fuel tank that created a flammable fuel-air mixture, as well as a better awareness of the myriad ignition sources present, and we were investigating various means of mitigating that risk. We knew that certain combinations of fuel loading and temperature would create such a flammable mixture, until the aircraft climbed into colder air and the fuel tank ullage, or vapor space, had cooled. We knew that this would obviously affect airplanes flown in the Middle East or American Southwest more so than those flown in Scandinavia or Alaska. The NTSB had already issued an Emergency Recommendation that argued for retroactively installed fuel tank inerting systems, but we were well aware of the problems associated with actually executing that recommendation, not only because at the time there was no commercially viable method for inerting a fuel tank, but because we knew that mandating the installation of a retrofitted system on the existing commercial fleet, as the NTSB was demanding, would be cost prohibitive to major segments of the industry.

As a pilot, I was a specific risk guy. All of my training had been oriented around spotting the lion, running from the lion, or, if need be, kicking the lion in the teeth. The horsefly problem was completely foreign to me. I was aiming at zero probability for individual airplanes in all climates, either by eliminating ignition sources altogether or by controlling the flammability of the ullage, shown as Condition B in Figure 4.2. Anything that allowed some aircraft operating in certain parts of the world to be more exposed to this threat than those aircraft based in colder climes seemed to miss the mark, to my way of thinking. As far as I was concerned, we needed to zero out the exposure, regardless of the environment.

If the probability of an ignition source = 1:

Condition A (Liquid Fuel is Non-flammable): Probability of an unsafe condition = 0 x 1 = 0

Condition B (Ullage is Either Too Rich or Too Lean): Probability of an unsafe condition = 0 x 1 = 0

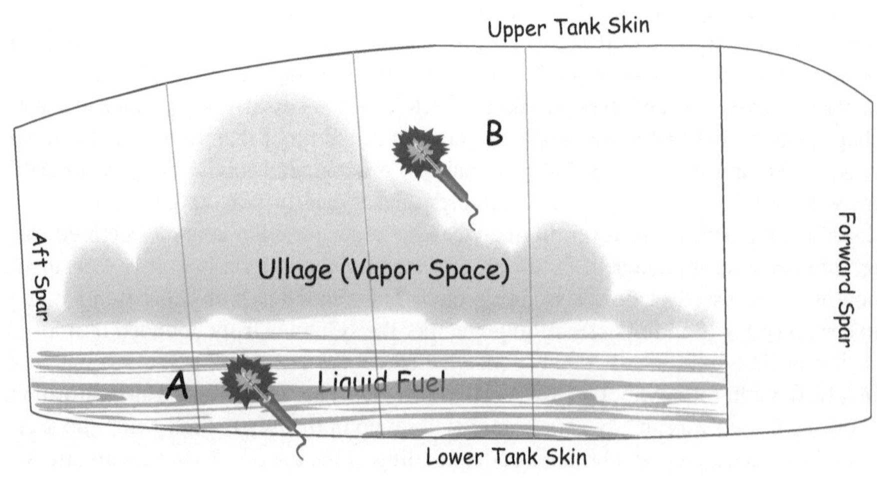

FIGURE 4.2 Under what conditions is the specific risk of a fuel tank ignition equal to zero?

This was a very tense time. There were thousands of airplanes operating all day every day all over the world. We had no way of knowing when the next fuel tank explosion might occur; we were struggling between the perfect solution that the Safety Board was strongly and publicly advocating, and a series of less-than-perfect approaches that we thought we could implement in a shorter time period … coincidentally, at less cost. It did not help that my understanding of risk was limited to lions.

Employing respectable skills of patient diplomacy while struggling to suppress his exasperation, the fuel guy from Boeing, a gentleman by the name of Ivor Thomas, explained that the fuel tank environments encountered by commercial aircraft around the world were too widely varied to control on a case-by-case basis. There would always be a fuel tank, somewhere in the world, that was operating in the flammable range, and thus exposed to the possibility of Condition C in Figure 4.3, should an ignition source be present. What he wanted to do was develop means of modifying the environment within all fuel tanks such that the period of flammability was reduced across the entire fleet. The industry was looking at things like directed ventilation and improved insulation to slow the heat transfer from existing air conditioning systems. This would reduce the exposure to the threat without attempting to change the already infinitesimal numbers of potential ignition sources … infected horseflies … that we believed existed across the entire world's commercial fleet. In

If the probability of an ignition source = 1:

Condition C: Probability of an unsafe condition = 1 (Ullage is a Flammable Mixture) x 1 = 1

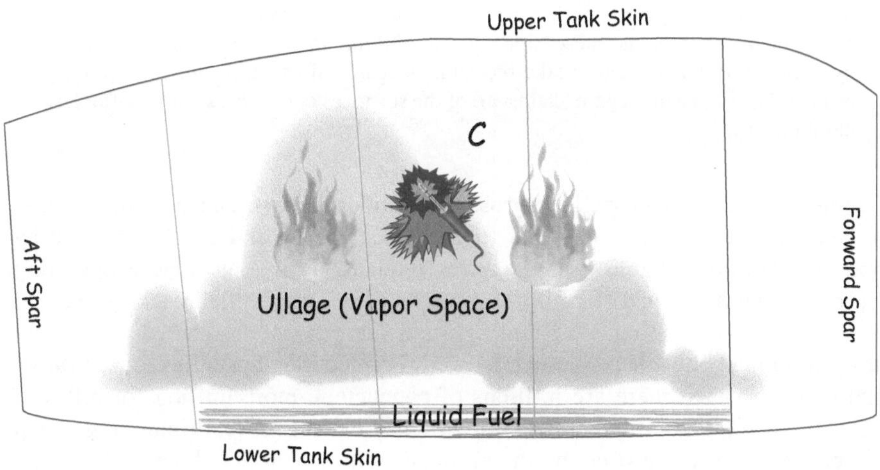

FIGURE 4.3 Under what conditions is the specific risk of a fuel tank ignition equal to one?

other words, it would reduce the averaged estimated probabilities of all unknown fuel tank events, but some would be reduced more than others. The numbers involved, thousands upon thousands of hours flown by thousands of airplanes in all regions of the world, precluded specific certainty on every flight with the technology that we thought we could lay our hands on. The entire discussion, at the time, revolved around the management of average risk.

By way of a simple example, the US Navy had already done this. Saddled with enormous amounts of jet fuel sloshing about in the lower bowels of an aircraft carrier, and with highly risky refueling operations going on above deck all day long, they tired quickly of the threat posed by the standard cut of jet fuel back in the day, which was JP4, the military's equivalent to Jet B. This is a mix of kerosene and gasoline. It has a pretty low flash point, the lowest temperature at which the fuel will ignite. This leads to all sorts of ramp and maintenance personnel setting new records for short distance sprinting, which is fine unless you happen to be on an aircraft carrier. JP4 can be difficult to handle safely.

The US Air Force also grew tired of blowing up tankers parked on the ramp in the middle of North Dakota, and opted to standardize their fuel on JP8, which is their equivalent to Jet A. Although it has a much higher flash point, the Navy still didn't like the degree of average risk mitigation that could be obtained using JP8. It wasn't just a matter of losing an airplane; they could lose the whole carrier. They settled on a cut of jet fuel known as JP5, which has a flash point twenty-four degrees higher than even JP8 does.[12] This, of course, drives their average risk way down. It is expensive stuff and it has its own problems … you need quite a bit of ignition energy just to get your airplane started … but the Navy prefers those problems over the threat of a major fire at sea.

At this point, we should step back and revisit St. Thomas' discussion of caution.

> Of the evils [threats] which man has to avoid, some are of frequent occurrence; the like can be grasped by reason, and against them caution is directed, either that they may be avoided altogether, or that they may do less harm. Others there are that occur rarely and by chance, and these, since they are infinite in number, cannot be grasped by reason, nor is man able to take precautions against them, although by exercising prudence he is able to prepare against all of the surprises of chance, so as to suffer less harm thereby.[13]

Cabin fires, runaway propellers, gross navigational errors, and lion attacks are, or were at one time, of somewhat frequent occurrence and can be grasped by reason. Certainly, engine failures were frequent; during my tenure operating big radial engines, I had six failures, with another seven pulled back to reduced power to keep it running. Fuel tank ignition sources such as chafed connectors and carbonized Kapton insulation, as well as infectious horseflies, occur rarely and by chance, but there are millions of connectors, more millions of miles of aircraft wiring, and billions of horseflies. They cannot (in real time) be grasped by reason … so we must define the prudence with which St. Thomas claims we can use to prepare against all, or at least most, of the surprises of chance, so as to suffer less harm thereby.

Condition D: Probability of an unsafe condition =

Estimated Probability of Ignition Source

x Estimated Probability of Flammable Vapor/Oxygen Ratio
(Considering the Liquid Surface Area, Temperature and Atmospheric Pressure at any given time)

x Estimated Probability of Temperature within the Flammable Range

= Some number between 0 and 1

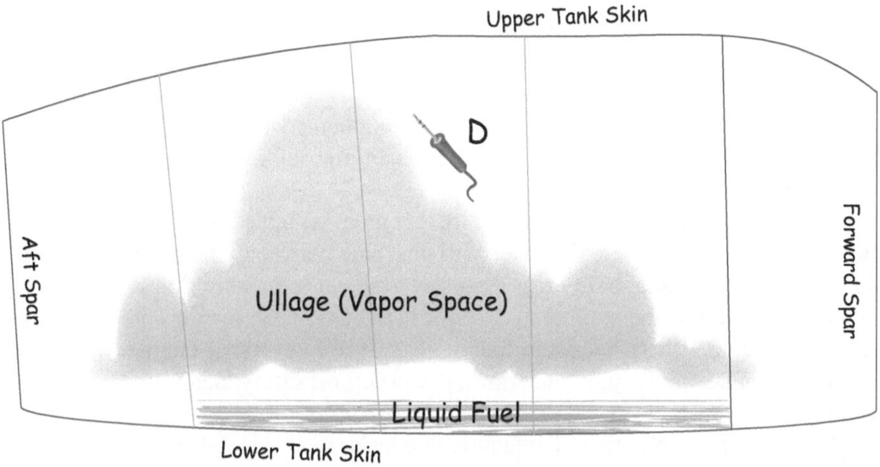

FIGURE 4.4 The probabilistic risk of a fuel tank ignition.

This is the kind of thinking that Ivor Thomas was trying to impress on me (although he did not go so far as to invoke the Summa Theologica). He was aiming at finding ways of minimizing the variables shown in Figure 4.4. In the end, we came at this from both directions; fuel tank environment and ignition sources. The FAA was not satisfied with a probabilistic risk approach that was limited to the fuel tank environment alone; they wanted to see how many potential ignition sources there were in the world. When they started looking, they, we, all of us, were shocked at what we found. Fuel pump shafts blued from overheating … high voltage conduit burned through from electrical arcing … it turned out that there were more than a few infected horseflies. The fleet average risk of a fuel tank ignition event was miles higher than we had initially thought. So, in addition to new designs employing fuel tank inerting systems, the design and inspection standards for anything that might represent an ignition source within a fuel tank were changed substantially with SFAR 88. Yet this, too, represented a reduction in probabilistic risk. We still have no way of knowing whether one specific fuel tank in one specific airplane has a chafed electrical connector exposed to the ullage. But we do know that the probability of that condition is significantly reduced through the use of updated inspection procedures and manufacturing processes.

THE RISK EQUATION

In structural engineering work, it is customary to apply a factor of safety to the load-bearing capabilities of structural components. By way of the analogy with horseflies and zebra stripes, *Marks' Standard Handbook for Mechanical Engineers* explains the process of selecting a safety factor by emphasizing that all uncertainties must be considered. The handbook elaborates to include magnitude and kind of operating load, material reliability, assumptions used in the applicable theories, the operating environment, localized stresses, stresses induced during fabrication, and uncertainty regarding possible causes of failure. They place particular emphasis on the possibility of failure *endangering human life.*[14]

If you know the strength of the material with some precision and the loads it will carry are steady and predictable, you might only apply a safety factor of three. This means that the structural member is capable of carrying three times the expected steady load without failing. The safety factor is planned for because of an unsafe condition resulting from the estimated probabilities of all unknown events … average risk.

If the loads applied to the structure become variable, the safety factor is increased to at least four. If the loads will alternate between tension and compression, the safety factor should be at least six. You can see where this is going. The less we know about the estimated probabilities of all unknown events, the fatter the safety factor. When you get to the crankshaft of a reversing engine, where the loads are complex and highly uncertain, the structural safety factor may be as high as forty.[15]

The easiest probabilistic risk equation to consider is a general form of quantifying risk. Figure 4.5 shows the elements of the equation and their range of possibility. From this, some obvious controls are evident. If, for example, we drive the exposure term to zero, perhaps by following Wilbur Wright's advice to sit on the fence, the risk is zero. If we reduce the severity, the risk is reduced. And if we reduce the probability of a single event, the risk is reduced.

Back in the 1980s, I frequently operated in and out of Baltimore-Washington International Airport in a Fairchild Metro. The runway pattern there is essentially an X shape. It was common for the controller to clear us to land and hold short of the

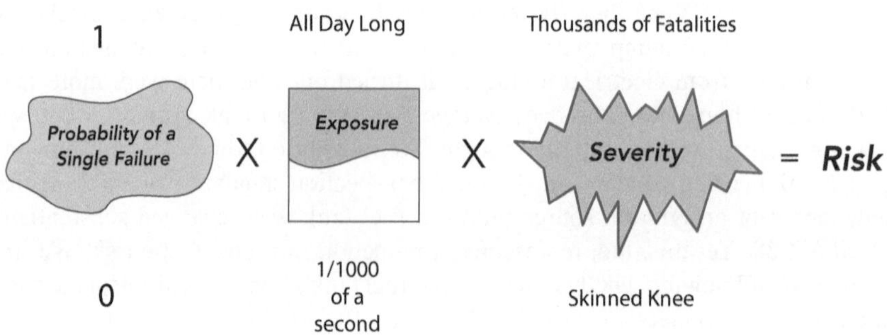

FIGURE 4.5 The basic risk model.

intersection, which in some cases gave us about 4,000 feet of runway. I refused this restriction; although we could land on a 4,000-foot runway, the risk was different. The single event I worried about was something like a longer-than-expected flare followed by a brake failure. This is an event with a low probability in either case. When landing at Baltimore and holding short, the probability of a brake failure AND entering the runway intersection at precisely the same time as the crossing traffic was considerably lower still. However, the severity was miles higher. If the brakes failed when landing at New Bern, for example, we would probably go off the end of the runway at twenty knots. Somebody would get a bruised knee. If they failed at BWI, on the other hand, and the crossing traffic was in the intersection at the same time … everybody would die. The difference in severity drove my decision to reject a hold short clearance.

In today's world, land-and-hold-short operations are generally prohibited if the runway is contaminated, ceiling or visibility is low, wind shear has been reported, or there is much of a tailwind. These are all probabilities that factor into the effect of something like a brake failure or thrust reverser failure, as well as the probability of both airplanes executing a go-around at the same time. Despite a low probability of this single event, the severity term is of such a magnitude that the risk equation must be mitigated. By codifying such restrictions into standard operating procedures, the exposure term is reduced, and that achieves the necessary risk mitigation.

PROBABILISTIC RISK ASSESSMENTS

Contemporary consideration of average risk uses a tool known as probabilistic risk assessment (PRA) to estimate the probabilities of all of the inputs and produce a probability of an ullage ignition, structural failure, loss of separation, or a fatal horsefly bite. This is a means of setting the table for the comprehension of uncertainty through the integration of historical failure data and physical limitations of components, equipment, and systems. A good general definition is provided by the Environmental Protection Agency's Risk Assessment Forum White Paper: "Probabilistic Risk Assessment Methods and Case Studies":

> Probabilistic risk assessment (PRA), in its simplest form, is a group of techniques that incorporate uncertainty and variability into risk assessments. Variability refers to the inherent natural variation, diversity and heterogeneity across time, space or individuals within a population or lifestage, while uncertainty refers to imperfect knowledge or a lack of precise knowledge of the physical world, either for specific values of interest or in the description of the system. Variability and uncertainty have the potential to result in overestimates or underestimates of the predicted risk.[16]

For the zebra, a PRA of a fatal horsefly bite involves very large numbers of inputs; the number of flies, how many carry disease, the zebra's skin temperature, the strength of the ammonia odor, the wind, and a host of others. These inputs to the equation are difficult to comprehend deterministically because they are random on any given day. However, their probability of occurrence over time can be estimated based on past history and carefully selected assumptions of randomization.

In a short paper on the topic titled "Probabilistic Risk Assessment: What Is It And Why Is It Worth Performing It?", Dr. Michael Stamatelatos of the NASA Office of Safety and Mission Assurance explains it this way:

> In a quantitative risk assessment or a probabilistic risk assessment, consequences are expressed numerically (e.g., the number of people potentially hurt or killed) and their likelihoods of occurrence are expressed as probabilities or frequencies (i.e., the number of occurrences or the probability of occurrence per unit time).
>
> Probabilistic Risk Assessment usually answers three basic questions:
>
> - What can go wrong with the studied technological entity, or what are the initiators or initiating events (undesirable starting events) that lead to adverse consequence(s)?
> - What and how severe are the potential detriments, or the adverse consequences that the technological entity may be eventually subjected to as a result of the occurrence of the initiator?
> - How likely to occur are these undesirable consequences, or what are their probabilities or frequencies?[17]

PRA is a way of managing uncertainty, whereas on the other hand, the control of specific risk, and the risk aversion buried within, is motivated by a drive to ensure certainty. Indeed, specific risk is predicated on the idea that we can see, and have seen, all of the ones coming that might get us. Probabilistic risk is a way of thinking about my core assertion, the idea that we will never actually see the one coming that gets us.

With the advent of nitrogen gas inerting systems that can be retrofitted onto existing designs, the FAA has come full circle back to an emphasis on specific risk with respect to fuel tank safety. In their disposition of Boeing's argument against retrofitting an NGS on the 757 aircraft, the FAA said that

> An acceptable average risk level and acceptable total fleet risk do not ensure that all reasonably anticipated flights (flights with known inoperative equipment, flights with undetected failures, flights in less-than-ideal but approved and expected weather or operational conditions, etc.) will provide the minimum level of safety expected by the public.[18]

This is an enormously important statement, another real foot stomper. The phrase "reasonably anticipated flights" actually means "reasonably anticipated stories", and is a direct reference to specific risk, unsafe conditions with a probability of one. When the true value or values of elements within the system are either known, or may be assumed because of service history, where the evils (threats) which man has to avoid *are* grasped by reason, a fleet average risk assessment is inadequate. You simply cannot integrate a data point with a probability of one into a Gaussian probability distribution, because the entire area under the bell curve is comprised of data points with a probability of between zero and one. More importantly, there is no place under the bell curve for a probability of zero.

The public is clearly averse to an inflight disintegration of the airplane, and in order to establish criteria that respect that risk aversion and mitigate the specific risk, it becomes necessary to assume that a specific flight operating with an undetected failure within the center fuel tank *will* experience an ullage explosion and the catastrophic destruction of the airframe. Likewise, a center fuel tank that operates one flight on a hot day in Phoenix with fifty gallons of fuel in an 85,000-pound tank *will* experience an ullage explosion.

In theory, probabilistic approaches to risk management can be combined with specific risk "worst-case" constraints, which is what the FAA has done in the fuel tank case by combining the ignition source reduction through SFAR 88[19] and requiring inerting systems as well. Likewise, regardless of whether you can meet the target for in-flight shutdown rates under an ETOPS program, you still train for engine failures. In practice, even these specific risk mitigations function probabilistically; the only way you can truly arrive at zero risk is to drive the exposure term to zero, usually by sitting on the fence. The nitrogen inerting system can be deferred under the Minimum Equipment List, and this is done based on the convolved probabilities of both an inoperative inerting system *and* the failures described in the FAA's disposition statement. This always reminds me of the old hitchhiking joke ... the hitchhiker asks, "how do you know that I'm not a serial killer?", to which the driver answers, "Because the probability of there being two serial killers in the same car is infintesimal!".

With that all said, let's look at a couple of examples in which the relationships between probabilistic risk and specific risk get muddled, convoluted, and, frankly, misused.

EXAMPLES

Flight dispatch programs at established airlines are very sophisticated and precise. Each engine's fuel consumption is tracked relentlessly, and the winds aloft forecasts have been pretty much spot-on for my entire career. The running two-year average holding times for each city pair are rigorously tracked. Reserve fuel is carefully indexed to aircraft gross weight. The operator has an obligation to minimize expenditure, and carrying more fuel than is necessary simply costs more fuel. The operator, when establishing policy for the automation of thousands of daily flight plans, has no choice but to use a PRA because there is simply no other way to do it.

And yet, inexperience with probabilistic risk can mesmerize the management folks into believing that the tails of a Gaussian distribution actually represent natural bounds to the system. On one memorable recurrent training slide, I recall the statement that the flight planning system was so good that less than 0.2% of the total annual flights operated by the airline would encroach on their FAA minimum reserve fuel. While factual, the construction of these statements is clearly oriented toward a suggestion that the company fuel plan rarely, if ever, need adjusting, because the statistical probabilities have all been addressed. Indeed, I once flew with a check airman who was quite proud of his claim that he had only added fuel once in his career.

The statement on the recurrent training slide is a rather specious argument made with no understanding of the relationship between specific risk and probabilistic risk, as described by the FAA's disposition of the Boeing 757 inerting retrofit AD. It comes down to the term, "reasonably anticipated flight". At the probabilistic level of automated fuel planning policy, the two-year average holding time for a particular city pair is certainly a pretty good way to "reasonably anticipate" a given flight. Certainly, the remarkably accurate fuel burn data and winds aloft data can be said to accurately define a "reasonably anticipated flight". But the open question is … at what point are you applying the term "anticipated"? A year ago, or today … right now … at the gate, before boarding?

For example, coming up the eastern Caribbean from Barbados to Miami in summer, a careful read of the details in the Miami weather forecaster's notes as well as consideration of the lapse rate, precipitable water, stability, wind shear, lifted index, and the like will give you a pretty good sense of the convective possibilities, regardless of the TAF. Some of the lines of storms building to the windward of Hispaniola and Cuba can become devilishly closely spaced, making the trip around the end somewhat lengthy. The problem is that if your fuel plan doesn't work out, the options for diversion are somewhat limited. Approaching the Florida coast, you can end up looking at a wall of storms with, hopefully, Nassau on your right, lots of exotic Cuban runways on your left, and Providenciales a good way back the way you came. Another twenty to thirty minutes of fuel on those days can make the entire three-and-a-half-hour flight much more comfortable.

The company is "reasonably anticipating" flights on a probabilistic basis months in advance. The dispatcher has a much better view of specific risks, the risks to any single flight, on a given day. In my experience, most dispatchers add an appropriate amount of fuel based on their analysis of that flight's probabilistic risk … the probability of a thunderstorm developing or visibility failing to come up … but he or she is also working fifteen or more flights at the same time. You, on the other hand, are working exactly one.

Moreover, there are aspects of probabilistic risk that one could easily argue the company is not "reasonably anticipating". On two occasions, I have dropped the wheels at the outer marker only to find less than the required number of little green lights. In both cases, we simply had a burned-out bulb; however, a go-around was required to be sure of that and to replace the bulb. The first one cost 1,100 pounds of fuel, the second 1,500 pounds. Had either situation involved an actual gear extension failure, arriving with the company minimums would have afforded us very little time to execute the appropriate QRH procedure, as well as almost no time at all to prepare the cabin for a real-live emergency landing. The aforementioned check pilot who had only added fuel once argued that the vast majority of pilots would never see such a situation, and he is correct. This is a probabilistic risk approach, and the industry is claiming that they have accounted for all of the specific risk involved, rendering the remaining average risk very, very low. Generally speaking, they are overestimating their skill.

I explained to my check airman friend that it only takes once. The severity term of the equation … 160 dead passengers … will not tolerate a mistake in the calculated probability of a single event. The probability of an actual gear extension failure is way lower than the probability of the hold time exceeding the two-year average. But the severity is miles higher. If the holding time exceeds my fuel load, I simply divert. If the wheels don't come down, I have a much bigger threat to manage.

Consider a second case. Civil Aeronautics Regulation Part 61, beginning in 1942, contained the well-known requirement that an air carrier aircraft must be dispatched at a weight that would allow it to land within 60% of the available runway length at the destination.[20] This requirement, like so many, has persevered down through the years and remains codified within FAR 121.195.[21] In 1990, Advisory Circular 121-195(d)-1A expounded on the variables that this provision was intended to protect against. These included "runway surface conditions, piloting techniques, tire and brake deterioration, atmospheric instability such as gusts or windshear, crosswinds, approach to touchdown, flightpath deviations, and others".[22] Although there is some specificity here, things like "piloting techniques" and "atmospheric instability" behave randomly in the real world, so this is a margin intended to protect against average risk or the estimated probabilities of all unknown events. The intention is to provide a 67% safety margin beyond the shortest landing distance demonstrated in flight test, which is a pretty reasonable guess at a safety factor for dry runways, a safety factor that would easily meet the criteria offered in Marks Standard Handbook for Mechanical Engineers. This is nothing less than the prudence that St. Thomas argued for.

The same type of calculation applied a 92% margin over the minimum demonstrated dry runway landing distance to situations in which the actual runway was wet. However, in recent years, it has become clear that the 92% margin has not provided an adequate margin against particularly slippery braking conditions. The authorities in Europe and the United States have promulgated a new approach to these less-than-optimum braking conditions, through an initiative known as TALPA, or the Takeoff and Landing Performance Assessment.[23] This approach takes into consideration a range of specific factors, such as a standardized description of runway conditions, additives to approach speed, runway slope, temperature, and others to calculate an actual landing distance; in particular, it generally assumes the use of reverse thrust, which the dispatch calculations assume to be inoperative. To this more precise landing distance is added a minimum 15% margin, a rather smaller margin than the old 67% and 92%. Essentially, this represents a paradigm shift of risk, placing considerably more weight on the control of specific risk and rather less weight on the margins used in average risk. Figure 4.6 shows how these margins compare with one another on a 7,000-foot runway.

As risk management becomes more and more quantitative and less qualitative, more of this type of transition from probabilistic to specific risk might be expected. As the landing overrun problem has proven, the consideration of unweighted average risk can be fallible, and the inclusion of more precise weighting of the probabilities, as well as the integration of elements of specific risk, is warranted. Going back to the engine with the reversing crankshaft, if you can very reliably calculate the maximum loads that the crankshaft may experience, you probably don't need a safety factor of forty.

But there are clearly economic drivers in play here as well. If you take the medium braking action raw distance for a typical twin jet transport and crank in the 92% safety factor, you end up with a number that exceeds the vast majority of runway lengths that a twin jet transport is likely to have been purchased to serve. We can see the door is wide open to the organizational and social forces that influence Amalberti's envelope of possible actions.

And yet, the ramifications of this paradigm shift from probabilistic to specific risk are poorly understood. For many, many years, the industry standard for landing

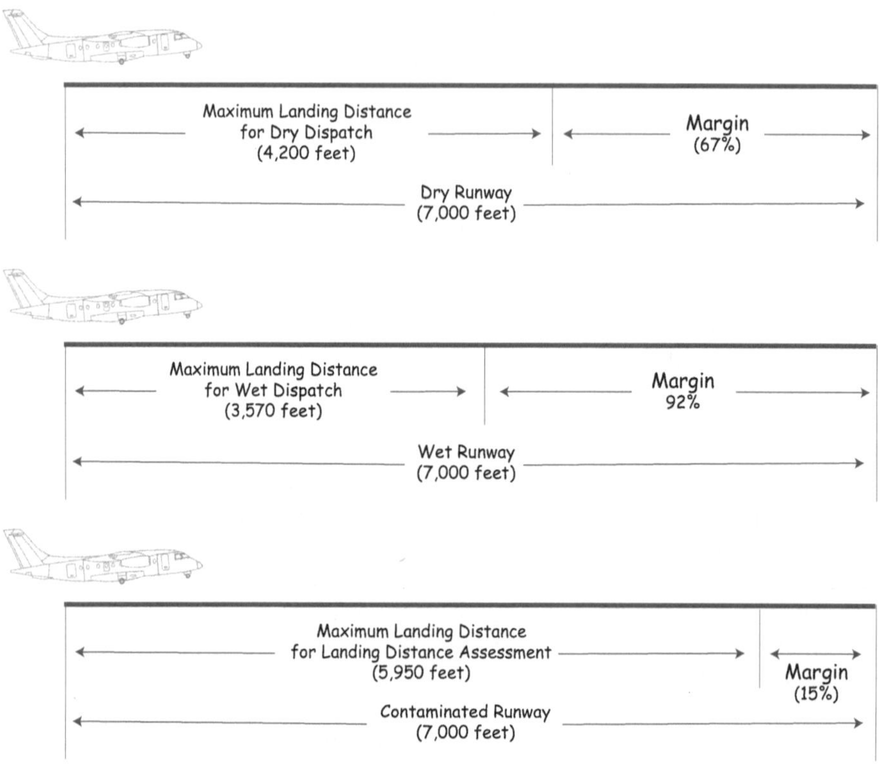

FIGURE 4.6 Comparison of acceptable stopping distances and margins for a 7,000-foot runway.

criteria has been to touch down within the touchdown zone, defined as the first third of the runway, or the first 3,000 feet for longer runways. What may go unnoticed is that the newly promulgated landing distance calculations are based on seven seconds of air time between the runway threshold and actual touchdown. That typically equates to between 1,400 and 1,700 feet. The distance between that number and the end of the touchdown zone is then *all or a substantial portion of the 15% margin.* Anytime you are actually relying on a landing distance calculation … slippery runways, short runways, etc., the standard policy of touching down within the touchdown zone is inadequate.

With the advent of TALPA, the industry has quantified many variables associated with landing distance, not the least of which is the runway surface condition. This is now expressed as the RCC, or Runway Condition Code. The actual number is generated by the airport authority using a range of inputs, and it truly is a huge improvement over the haphazard approach used in the past. That said, the RCC is also directly correlated with the landing distance and its 15% margin. The idea, as with many of these schemes, is that we can more precisely calculate the unique, situationally dependent landing distance using a better characterization of the runway condition. And, so we can … until the RCC is dramatically altered while we are on short final.

Approaching Minneapolis one night, it was snowing. Again. The ATIS was advertising the RCC for 12R as a 5, equivalent to a wet runway with no standing water, and correlated with good braking action. The first officer pulled up our new handy-dandy landing distance app on his iPad, worked out the correct landing distance as required, and we were all set. Intercepting the localizer, totally immersed in cloud, the controller happened to mention that the preceding aircraft reported the braking action as fair, which, in the book, corresponds to an RCC of 3, which is not what we had calculated.

Sure enough, in 2019, the FAA issued SAFO 19003, Turbojet Braking Performance on Wet Runways. The letter stated that

> Several recent runway-landing incidents/accidents have raised concerns with wet runway stopping performance assumptions. Analysis of the stopping data from these incidents/accidents indicates the braking coefficient of friction in each case was significantly lower than expected for a wet runway as defined by Title 14 of the Code of Federal Regulations (14 CFR) part 25 § Section 25.109 and Advisory Circular (AC) 25-7D methods.[24]

The SAFO went on to state that "The data indicates that applying a 15% safety margin to wet runway time-of-arrival advisory data, as recommended by SAFO 19001 (or current guidance), may be inadequate in certain wet runway conditions".[25]

The problem is that, buried within the RCAM, there is a subtle yet powerful change with respect to the depth of water on the runway. If the water depth is 1/8 inch or less, an RCC of 5 is acceptable. However, if the water depth is greater than 1/8 inch, the RCC changes to 2, a rather substantial change that can lead to a need for several thousand additional feet of runway length. Nobody is going to go out and check the water depth at various points on the runway every few minutes; the FAA pointed out in SAFO 19003 that "Rainfall intensity may be the only indication available to the pilot that the water depth present on the runway may be excessive"[26] and that "30 to 40 percent of additional stopping distance may be required if the runway transitions from *wet* to *contaminated* based on the rainfall intensity or reported water contamination (greater than 1/8-inch depth)".[27]

So much for a precise accounting of specific risks.

In order to successfully forage, the zebra really has no way of controlling his exposure to either lions or horseflies, aside from not hanging around the water hole too long. Natural selection has instead perpetuated a unique means of reducing the probability of a single event through the dominance of a genetic pattern leading to stripes. In both zebra-stripe theories, the stripes operate as a passive system of threat mitigation, reducing the probability of a successful lion attack or the probability of a fatal horsefly bite without any active input from the zebra. The zebra doesn't have to be better at lion detection, doesn't have to smell better, listen more closely, or watch more keenly. The stripes effectively disrupt the lion's plan of attack by virtue of simply being there, reducing the likelihood of a successful attack, and that probably prevents fatter lions. Likewise, the total number of horsefly bites on any given day is substantially reduced, because, apparently, there are simply fewer horseflies willing to land on a striped surface. This effectively reduces both the probability of a critical blood loss and the probability of infection. The stripes are a consistent defense against an undetected, unseen, or unanticipated threat, a defense that does not depend on individual skill or state of mind. They are not a guarantee and thus are not effective at resolving any particular degree of risk aversion. Rather, they are a defense against the convolved average risk of death from both lions and horseflies. They simply reduce the probabilities, mitigating uncertainty.

But the question remains: how confident are you that you have captured all of the risk? Clearly, a zebra in a zoo doesn't need stripes to protect him against lions. But the horseflies remain. As the pilot-in-command, our job is to protect the margins. It is also to practice what Karl Weick has referred to as a sensitivity to operations.[28] On any given day, a thorough review of the weather forecasts, trends, deferred equipment, and NOTAMs will allow the pilot to identify unique cases of specific risk that are applicable to that flight on that day. Regardless of how well-fine-tuned the company's fuel burn is, on some days, the margin needs to be fatter. And no matter how precisely you have entered the landing distance inputs into your little iPad app, on most days when it is truly applicable, a 15% landing distance margin is very likely to be inadequate.

And sure enough, there are still those black swans. The years since World War II have built a pile of data, which paved the way for a wide range of alchemists to mangle the language of Bruner's paradigmatic mode of thought while attempting to turn a black swan into a common duck. But that begs the question; a black swan is not a specific risk, a known known, so-to-speak, and it isn't an average risk, either ... a known unknown. Rather, it is an unknown unknown. There is no estimated probability; it isn't even on the radar. That's the point. It can't be defended against in the sense of a feedback-driven intervention. Like the zebra successfully hunted by the lioness, you simply will not see it coming; but at that point it departs from the "known known" threat posed by the lioness. The lioness is unseen largely due to the natural limitations in the zebra's sensory data acquisition and/or cognitive processing of that data, but she is not unexpected. The black swan is unexpected because nothing in the past can clearly point to its possibility. You cannot think of a thought until you think of it, but when you do arrive at the thought of a black swan, you will dismiss it out of hand. Up until a particular point in time, which is always too late, all the swans in your experience will have been white.

On October 31, 1994, Simmons 4184, an ATR-72, encountered an aerodynamic phenomenon known as aileron snatch during a descent in icing conditions.[29] In this situation, ice builds up behind the protected surface, and generates a flow separation bubble that migrates aft over the flight control hinge balance point. The aileron, if it is not a powered, irreversible control, becomes aerodynamically imbalanced but remains effective in roll. The crew experiences a rapid and violent oscillatory displacement of the control wheel and a resulting roll force which cannot be reversed as long as the angle of attack remains above whatever angle allows the separation bubble to reach the flight control balance point. It is quite fair to say that, for this crew as well as nearly everyone in commercial aviation at the time, nothing in their previous experience "could convincingly point to its possibility". Likewise, when the first officer of American 585 applied significant reversing rudder inputs as a technique for overcoming what he perceived as a roll upset, nothing in his experience could convincingly point to the possibility that the vertical stabilizer would actually depart the airframe.[30]

Yet, in both the Simmons and the American accidents, the mechanics of causation were not outside of the comprehensive body of aeronautical engineering experience. Rather, the complete set of available data used in assessing these risks … at least at the point of original certification of the design … yielded no reason to construct such scenarios. Part of the problem may lie in the use of "normal" bell-curve probability distributions when analyzing probabilistic data that is historically weighted.

THIN TAILS AND FAT TAILS

A serious mistake can be made if we assume that the thin tails of the bell-shaped curve actually define a bound. Let's consider a comparison of individual human height vs. individual net worth. If we randomly select two people, and add up their heights, we get some number, say twelve feet. What is the probability that one of them is ten feet tall and one is two feet tall? Zero. What is the most likely pair of two possible heights? Somewhere in the range of 5'10" to 6'2" … we know that we can use a Gaussian distribution with this, because the physical limits of human height represent a worst-case boundary.

Now suppose we add two people's net worth together, and we get a total of 40 billion dollars. What is the most likely combination of individual net worth that will yield such a result? One billion and 39 billion? Ten billion and 30 billion? Obviously not. The most likely combination is a few thousand dollars and 40 billion minus a few thousand dollars. Hidden in this case is the fact that a Gaussian probability distribution of human wealth will reveal some reasonable average, say ten thousand dollars, with the tails of the bell curve extending out approximately three standard deviations from that average. The average personal net worth will be fairly close to what we might expect, because nearly everybody on the planet has a very modest net worth. The guy with 40 billion dollars will not even be on the same page; he will be excluded as an outlier, because his wealth is so far beyond the vast bulk of the rest of humanity. But his decisions and actions matter greatly to the world economy.

Here we begin to get a sense of where we might find black swans. Theoretical statistics operates under the premise that a relatively small sample size will mimic the distribution of all possible data; we're back to the Central Limit Theorem. But unbounded probabilities require enormous amounts of data to accurately establish a true distribution. For practical purposes, it is much easier, since we will never have enough data, to handle these situations as if they were normally distributed, and ignore the outliers. The mistake is in assuming that the thin tails of the bell curve represent an actual boundary. In fact, the guy with 40 billion exists under what is known as a fat-tailed distribution.

Taleb and others have spent several years beating the drum of the idea of fat-tailed probability distributions, which debunks and discards the idea of a standard deviation from the average as a measure of acceptable predictability and instead argues that the catastrophic outliers define the probabilities.[31] Rather than building a probability distribution based only on an average and the deviations from that average, a fat-tailed distribution is developed from what is known as a power law, which is a type of exponential relationship that more explicitly reveals the possibility of catastrophic outliers farther from the bulk of the sample size than a normal distribution would. In fact, a power law distribution implies a much greater probability of a catastrophic outlier than the normal distribution does.

Both the thin-tailed, Gaussian probability distribution and the fat-tailed, power law type of distribution have a role to play in managing uncertainty, but the role depends on how bounded the relationship is. TWA 800 was about as black a swan as you can get. A PRA of fuel tank safety would not have predicted it; we had, and really still have, nowhere near enough data on these events to accurately derive the correct exponent defining the power law relationship that yields the actual fat-tailed distribution. Indeed, during the early work of the Fuel Tank Harmonization Working Group, we arrived at a probability of another catastrophic fuel tank explosion occurring within a year or two. This has not happened, which would suggest that the probabilistic assessment with the tools and data we had was not accurate.

Recent work by Sornette et al., at ETH Zurich, has examined the probability of nuclear power plant catastrophes as predicted by standard probability risk assessment techniques as compared to their actual occurrences.[32] They found that the PRA approach predicted 0.1 releases of radioactive contaminants from the reactor building over the course of human history, not the two to four actual releases of contaminants that they documented from empirical data. They concluded that "In particular, PRA is currently not adapted to identify correctly the likelihood of large catastrophes".[33] Such is the effect of constraining the discussion to normal, thin-tailed distributions.

We reduce the probabilities of single events through the use of sophisticated design features such as the electrical bonding of the entire aircraft structure to reduce the probability of a lightning strike creating an arc within the fuel tanks, or hydraulic fuses (like the DC-10 did not have prior to the United 232 accident) to reduce the likelihood of a total loss of all hydraulic fluid from all of the hydraulic systems. We can reduce the probabilities of single events in the cockpit, as well as control the exposure to threats, through adherence to standard operating procedures.

The black swan is, by definition, an event that we cannot manage through risk aversive techniques centered around specific risk, or through PRAs using normal probability distributions. One approach to anticipating the possibility of a black swan is to build in redundancy. We can build barriers to black swans by constructing independent and alternative system paths to a successful outcome, which is the point of multiple pitot-static systems and powering the standby busses normally with the AC system and alternatively with the battery.

These techniques are how we create regions of resiliency, putting some distance between ourselves and a stalking lion, and generally discouraging horseflies. But the stripes alone clearly do not completely mitigate the risks; zebras are still killed by both lions and horsefly-borne disease. Simply sporting a set of snazzy stripes is no excuse for ignoring a whiff of a lioness or, for that matter, mindlessly standing around in the tall grass to begin with. A capable and competent zebra still practices a form of mindfulness, paying excruciatingly close attention while foraging.

In his work on high-reliability organizations, Karl Weick has offered a pocket-sized definition of mindfulness: it is, he says, a "rich awareness of discriminatory detail".[34] He elaborates on this by saying that being mindful means paying attention in a different way; it is to see more clearly, not to think harder and longer. You stop concentrating on those things that "confirm your hunches, are pleasant, feel certain, seem factual, are explicit, and that others agree on".[35] You start concentrating on things that "disconfirm, are unpleasant, feel uncertain, seem possible, are implicit, and are contested".[36]

While it is doubtful that our wise old zebra contemplates this point to quite such depth, all one needs to do is watch how fast his nostrils flare and his movements freeze the second he detects an odor that, so-to-speak, disconfirms his hunches. No doubt munching down the forage is pleasant, but that's not what he is concentrating on.

The application of a rich awareness in discriminatory detail, searching for things that disconfirm or are contested, can force the creation of additional margins, as well as provide ample opportunity to fatten the ones you already have. In some cases … not all … but in some cases, it can provide the means of looking beyond experience, possibly climbing the parapet, peering over the wall of experience, looking out into the vast unknown, and strangely enough spotting a couple of extra swans appearing in places that you would swear they hadn't been before.

The crew of Simmons 4184 had no warning, no unusual feel, no gut sense, that might have led them to reconsider their margins. They were aware of the ice but were also aware that the descent they were executing would change the icing environment, and they were operating all of the ice protection equipment correctly. It is reasonable to assume that they did not sense any buffet or buzz in the flight controls prior to the upset, perhaps due to a fairly rapid increase in angle of attack following flap retraction as they accelerated from the holding speed. But within two hours of their accident, an identical aircraft just a few miles away, at South Bend, Indiana, did experience an airframe buffet while descending for their approach. No ice was visible on any of the critical locations, including the leading edge of the wing. As with Simmons 4184, all ice protection systems were operating. The captain advanced power and accelerated, reducing the angle of attack … protecting the margins. The buffet stopped. He and his first officer then inspected the wing in more detail and discovered a ridge of ice aft of the protected surfaces. They made the decision to delay any configuration changes until they descended into warmer air and executed a normal landing. The captain said that he had never experienced ice accretion of this nature before.[37]

Both crews had operated the ice protection systems correctly. In so doing, they eliminated any number of unknown events, using procedure to remain within the space of safe operation by keeping the wing as clean as the system would allow. Only the South Bend crew was afforded the time to investigate the discriminatory detail, the buffet. Procedural compliance is a powerful tool within the paradigmatic mode of thought, preserving resilience in a world of average risk, up to the point that mindfulness intervenes with something that disconfirms or is uncomfortable. Mindfulness is a powerful tool within the narrative mode of thought. It begins with an acceptance of uncertainty, but is only effective if you do not misinterpret, and in particular overestimate, the value of your experience.

> But in all my experience, I have never been in any accident ... of any sort worth speaking about. I have seen but one vessel in distress in all my years at sea. I never saw a wreck and never have been wrecked nor was I ever in any predicament that threatened to end in disaster of any sort. Captain E.J. Smith, 1907[38]

In 1912, Captain Smith assumed command of the Titanic for its maiden voyage.

NOTES

1. Nicholas Nassim Taleb, *The Black Swan*, Random House, 2007, 165–200.
2. Ibid., 302.
3. Ibid., xvii.
4. Ibid., 40.
5. Ibid., 100.
6. ARAC ASAWG Report, *Specific Risk Tasking,* Federal Register, Vol. 71, No. 54, Tuesday, March 21, 2006.
7. Sam Pullen, Todd Walter, Per Enge, "SBAS and GBAS Integrity for Non-Aviation Users: Moving Away from 'Specific Risk'", International Technical Meeting of the Institute of Navigation, San Diego, CA, January 24–26, 2011.
8. Brenda Larison et al., "How the Zebra Got Its Stripes: A Problem with Too Many Solutions", *Royal Society Open Science*, Vol. 2, No. 1, 2015, 140452, doi: 10.1098/rsos.140452.
9. T Caro, et al., "Benefits of zebra stripes: Behaviour of Tabanid flies around zebras and horses", *PLoS ONE*, Vol. 14, No. 2, 2019, e0210831, doi: 10.1371/journal.pone.0210831.
10. United Nations Office for Disaster Risk Reduction, *Deterministic and Probabilistic Risk*, https://www.preventionweb.net/disaster-risk/concepts/deterministic-probabilistic/.
11. Pullen, "SBAS and GBAS Integrity...".
12. Skybrary, "Ignition of Fuels", https://skybrary.aero/articles/ignition-fuels, March 13, 2022.
13. Thomas Aquinas, *Summa Theologica*, translated by Fathers of the English Dominican Province, e-artnow Kindle Edition, 2013, IIa-IIae, q.49, a.8, ad.obj.3.
14. Theodore Baumeister, Editor, *Marks' Standard Handbook for Mechanical Engineers.* Seventh Edition, McGraw-Hill, Inc., 1967.
15. Engineers Edge, "Factor of Safety Review", https://www.engineersedge.com/analysis/factor-of-safety-review.htm, March 13, 2022.
16. United States Environmental Protection Agency, *Risk Assessment Forum White Paper: Probabilistic Risk Assessment Methods and Case Studies,* EPA/100/R-14/004, July 2014.
17. Dr. Michael Stamatelatos, Probabilistic Risk Assessment: "What Is It And Why Is It Worth Performing It?" NASA Office of Safety and Mission Assurance, April 2000.
18. Federal Register, Vol. 80, No. 35, Monday, February 23, 2015, Proposed Rules, Docket No. FAA–2012–0187; Airworthiness Directives; the Boeing Company Airplanes.
19. Federal Aviation Administration, 14 CFR Appendix Special Federal Aviation Regulation No. 88 – Fuel Tank System Fault Tolerance Evaluation Requirements, 2001.
20. United States Civil Air Regulations, Part 61, Scheduled Air Carrier Rules, October 15, 1942.
21. United States Code of Federal Regulation, Title 14, Subchapter G, Part 121, 1965.
22. Federal Aviation Administration, Advisory Circular 121-195-1A, "Operational Landing Distances for Wet Runways; Transport Category Airplanes", June, 1990.

23. Federal Aviation Administration, Advisory Circular 91-79A, Change 2, "Mitigating the Risks of a Runway Overrun Upon Landing", February, 2018.
24. Federal Aviation Administration, Safety Alert for Operators 19003, July 2, 2019, Flight Standards Service, Washington, D.C.
25. Ibid.
26. Ibid.
27. Ibid.
28. Karl Weick, Kathleen Sutcliffe, *Managing the Unexpected: Assuring High Performance in an Age of Complexity* Wiley, 2015. Kindle Edition, 77.
29. National Transportation Safety Board, AAR-96/01, "In-Flight Icing Encounter and Loss of Control Simmons Airlines, D.B.A. American Eagle Flight 4184 Avions De Transport Regional (ATR) Model 72-212, N401AM, Roselawn, Indiana October 31, 1994".
30. National Transportation Safety Board, AAR-04/04, "In-Flight Separation of Vertical Stabilizer American Airlines Flight 587 Airbus Industrie A300-605R, N14053 Belle Harbor, New York, November 12, 2001".
31. Taleb, *The Black Swan.*
32. D. Sornette et al., "Exploring the limits of safety analysis in complex technological systems", *International Journal of Disater Risk Reduction*, Vol. 6, December 2013, 59–66.
33. Ibid.
34. Karl Weick, "Reflections on Enacted Sensemaking in the Bhopal Disaster" *Journal of Management Studies*, Vol. 47, No. 3, May 2010, doi: 10.1111/j.1467-6486.2010.00900.x.
35. Karl Weick, Kathleen Sutcliffe, "Managing the Unexpected", 2nd Managing the Unexpected in Wildland Prescribed Fire Workshop, Jacksonville, Florida, February 28, 2005.
36. Ibid.
37. NASA Aviation Safety Reporting System, Report Number 287615, October 1994.
38. The Quotations Page, http://www.quotationspage.com/quote/22207.html, March 13, 2022.

5 Complexity

On October 7, 2008, a Qantas Airlines flight 72, an Airbus 330, experienced an inflight upset while approaching the coast of western Australia enroute from Singapore. The upset was initially manifested by the autopilot disconnecting, followed by a series of seemingly unrelated ECAM messages and both stall and overspeed warnings. The captain's primary flight display airspeed and altitude indications appeared erratic. Within a couple of minutes, the aircraft spontaneously pitched down to just over eight degrees nose down, incurring a load factor of negative .8 g and injuring several passengers. A second pitch-over occurred shortly afterward. The crew successfully landed at Learmonth, Australia, fifty-two minutes after the onset of the event.[1]

This event was ultimately traced to multiple spikes of data from the no. 1 ADIRU. There were two significant problems identified by the investigation; first, the multiple data spikes, and second, the failure of the FCPC design to prevent the erroneous data from being interpreted as valid. The investigation was quite lengthy and detailed. It is perhaps the first seriously deep dive into the processes used for SSA up to that point in the history of accident investigation. SSA processes incorporate such methodologies as the use of a fault tree analysis as well as a FMEA. The fault tree begins with a failure and looks backward to find causality. The FMEA begins with a failure as well and looks forward through the system to find out what the ramifications would be. Both are hugely subjective; the report on this incident went into great detail on the dependency of the entire SSA process on engineering judgment as well as the ability to anticipate possible failure scenarios. Engineering judgment is largely based on experience and has been responsible for much of the progress we have made in technology and safety assessment to date. However, as the report noted, there has been virtually no research done on the human factors of engineering judgment … how it is obtained and how it is applied … leaving the process of the SSA wide open itself to human error. And as far as the design engineers failing to anticipate particular failure scenarios, well, that's yet another case of "you can't think of a thought until you think of it".

The A330 has three separate angle-of-attack probes: one on the left side and two on the right side. A typical three-probe design would compare each individual probe input with the median of all three. If one probe's input deviates too far from the median, the data is rejected. However, during original flight testing of the A340, a problem with this algorithm cropped up. During a sideslip condition, it was possible for the two probes on the right side to be incorrect, while the left side probe was rejected for straying too far from the median value.

Airbus fixed this with a change to the algorithm. The check against the median value of AoA was retained; however, the actual angle-of-attack input to the FCPC was the average of AoA probes 1 and 2. If one of these two probes was the errant probe, showing a deviation from the median, then the last valid AoA from that probe was "memorized" for 1.2 seconds. Essentially, the FCPC closed its eyes and held its breath for 1.2 seconds, then opened its eyes to see what was what.

DOI: 10.1201/9781003369677-5

The problem in the new algorithm was that whatever the FCPC saw at the end of 1.2 seconds was taken as valid data, regardless of where it sat with respect to the median. While the 1.2-second assumption might be adequate for most errors due to sideslip condition in large jet transports, it is a very different story for corrupted data. The original data spike had indicated an angle-of-attack of 50.6 degrees. The FCPC took that to be erroneous, ignored it, and used a value of 2.3 degrees. The original data spike quickly disappeared … and was followed by another data spike of exactly the same magnitude, which was still present at the end of the 1.2-second memorization period. And *that* spike was interpreted as valid because of the flaw in the FCPC algorithm. And so it went.

What you've just read is a narrative, and it fits the template described by Bruner. It is a story, with the departure from the canonical order of things being the pitch down of the aircraft, and the rest of the information explaining how this happened forming the story that makes this event comprehensible. It is a linear analysis; the summing of the logical gates that the data passed through to cause the elevator deflection yields exactly what happened, a resultant value that, when recreated in simulation, does just that … pushes the nose over. So far, so good.

But how did we end up with a data spike in the AoA from the no. 1 ADIRU that happened to equal 50.6 degrees? And how did we end up with a second data spike of exactly the same magnitude less than 1.2 seconds later?

There were actually quite a number of data spikes, reflecting problems with AoA but also Mach, calibrated airspeed, and so forth. Most of these data spikes were rejected by the FCPC, but, as we have seen, a particular flaw in a particular piece of the code allowed some data to be taken as valid. The final report made a number of interesting observations, but there are three that I think are particularly significant:

- One of the aircraft's three ADIRUs exhibited a data-spike failure mode, during which it transmitted a significant amount of incorrect data on air data parameters to other aircraft systems, without flagging that this data was invalid.
- It is very likely that the ADIRU data-spike failure mode involved a problem with the data packaging and queuing within the ADIRU's central processing unit module. *This fault resulted in numerous data anomalies, including air data reference parameters being intermittently transmitted with the data or label of another parameter* (italics added). Despite extensive testing and analysis, the exact origins of the failure mode could not be determined.
- Tests and analyses showed that the ADIRU data-spike failure mode was probably not triggered by a software bug, software corruption, hardware fault, physical environment factors (such as temperature or vibration), or from electromagnetic interference.[2]

This particular ADIRU design had accumulated over 128 million flight hours and experienced the data-spike failure mode precisely three times.[3] Two of these occurrences were on the same aircraft within the same ADIRU. This particular occurrence was the only one of the three that resulted in uncommanded flight control inputs. And nothing in their collective and extensive previous experience allowed investigators to track a cause. It was nothing less than a card-carrying, registered, grade-A black swan.

The French mathematician, Pierre Laplace, once argued that the present state of the universe was the effect of its preceding state and the cause of the following state. He said that "Given for one instant an intelligence which could comprehend all the forces by which nature is animated and the respective situation of the beings who compose it …" one could comprehend everything from the movements of the planets to the movements of atoms; nothing would be uncertain.[4]

"An intelligence which could comprehend all the forces by which nature is animated" sounds rather similar to an "intellective power [that] is in full vigor", which can comprehend "their truth by their simple insight, as do God and the angels".[5] Statements such as these are the basis for the idea of Newtonian scientism that Sidney Dekker has fairly criticized, the idea that "… total knowledge of the world is achievable", and that "People will know the truth if they are fully rational, once the correspondence between the picture in their mind and reality in the world is perfect".[6]

The open question is whether Laplace's concept is expressed in the form that it is, or even exists at all, because of the human reliance on the narrative mode of thought. And if so, does such a construction of thought narrow our ability to comprehend what has happened? Because those ADIRU spikes are a problem.

Although they made a very comprehensive attempt to identify the fault that led to the ADIRU spikes, the investigators ran into some significant difficulties.

Failure mechanism – a series of analyses determined that the failure mode almost certainly occurred with the ADIRU's central processing unit (CPU) module. More specifically, evidence indicated that many of the data spikes for ADR parameters were produced when the CPU module packaged the 32-bit output data words. These data spikes were found to be the result of the data word being packaged with either the wrong label field, or the wrong data field. Evidence also indicated that the CPU module's data processing stages before and after the data packaging operated normally for the ADR data.

The packaging problem was intermittent rather than consistent. In addition, it was very unlikely that each data spike was due to a separate fault, failure or trigger event. A much more likely scenario is that *the failure initiated a problem in a higher-order process* for organizing the storage or retrieval of buffered data within the CPU module. Other symptoms of the failure mode, such as the corruption of the IR data and the storing of BITE data, also involved buffering data in memory. The investigation was not able to identify the precise mechanism involved, although some possibilities were able to be excluded (such as corruption of the data labels or problems with the wait states).

Unit susceptibility – as discussed above, the failure mode probably involved some form of hardware problem. The data-packaging process involved several components within the CPU module, including the CPU chip, application-specific integrated circuit (ASIC), a wait-state random access memory (RAM) chip, and four RAM chips for general CPU use. The investigation was not able to identify which of these component(s) were directly involved.

There can be variations in the properties of components with the same part number, including those manufactured within the same batch but even more so for those manufactured in different batches. The two affected ADIRUs had CPU-module components that were manufactured in the same or adjacent batches. Due to limitations in the way that component details were recorded, the investigation was unable to determine how many other ADIRUs contained components that were manufactured in the

> same or adjacent batches to these units. However, it was known that most of the units manufactured after these units (after the end of 2002) had a redesigned CPU module with some different components.[7]

Variations in the properties of components with the same part number appear to speak to an issue of manufacturing quality control, and from the point of view of Laplace or Newton, it sounds very much like the language that we may be familiar with from previous accident investigations involving issues of quality control, for example, the language used when investigating the presence of hard alpha metallurgical defects in titanium alloy ingots, from which turbine engine compressor disks are forged, as was the case with the United 232 accident at Sioux City.[8] Or perhaps the language used when investigating some form of drill breakdown, combined with localized loss of coolant and chip packing, like that which created an altered microstructure and ladder cracking in the left engine's fan hub of Delta 1288 at Pensacola.[9] Both of these accidents led to significant recommendations and changes in quality control processes. But note that the Qantas report simply states that there can be variations in the properties of components; it makes no specific recommendation regarding quality control (although the report did note that Northrop Grumman was evaluating methods to improve the robustness of the CPU). In fact, while the language is similar, these are not quite the same thing. The report noted that

> The ADIRU outputted numerous spikes on air data reference (ADR) parameters. The spikes had short durations (less than 1 second), occurred at different times and frequencies for each parameter, and had a limited number of values for many of the parameters. With the exception of the data spikes, all of the ADR output data appeared to be correct. The ADIRU outputted the data spikes to other systems as valid data.[10]

And, perhaps the most important aspect to this type of investigation, the ADIRU stopped generating data spikes after it was powered down and powered back up again, exhibiting what is known as a "soft" fault.[11] Such conditions lead to well-known maintenance logbook phraseology, such as "could not duplicate" and "ground checked okay". In contrast, hard alpha defects and ladder cracking don't disappear just because you shut the engine down. As such, we need to be careful not to approach the ADIRU problems in the same manner of thinking that we approached the fan disk problems, because while the behaviors of hard alpha inclusions or ladder cracking can be thought of as fairly linear cause-and-effect relationships, the behavior of CPU modules may not be linear at all.

NONLINEARITY

The concept of nonlinearity has a range of interpretations. The easiest way to understand it is to start with what we mean by linearity. A process or system is considered linear when the output of that process is proportional to the input. Basically, this means that we can draw a straight line on an XY axis plot, plug in a value for X and get a proportional value of Y. A quarter turn on the volume knob gives you a quarter turn's worth ... a proportional ... increase in volume.

For more sophisticated systems, we can consider them as linear when the superposition principle applies. This means that we can break the system up into its component parts, measure the output of each component, add them all up, and expect to get exactly the same output that we would have gotten had we left the whole thing alone. If we are certain that some function f(x+y+z) is always equal to f(x) + f(y) + f(z), then we are dealing with a linear system. Superposition gives us proportionality.

As pilots, we are very accustomed to linear response, because we depend on it when operating flight controls. For a given movement of the yoke, we expect a proportional response from the airplane. Interestingly enough, aerodynamics has a pretty strong aspect of nonlinearity associated with it. Turbulent flow, as opposed to laminar flow, is the Holy Grail of nonlinear problems, and the Navier–Stokes equations that are used to model aerodynamic flows are comprised of nonlinear partial differential equations. The challenge for the aerodynamicist is to produce a linear control response and stability behavior by optimizing nonlinear fluid flow behaviors.

Nowhere is this more clearly borne out than when considering the problems of inflight ice accretion. For about fifty years, the FAA published a little diagram showing a cartoon airplane accreting inflight ice, with the four forces appropriately degraded; less lift, more weight, more drag, less thrust. The diagram, reproduced in Figure 5.1, carried the statement, "The effects of icing are cumulative".[12] While linguistically correct, it is semantically misleading. Here is a classic misuse of linearity.

The word cumulative strongly implies linearity, encoding superposition by suggesting that proportional changes in degradations will add up to an output that is proportional to the sum of the degradations. That is not how icing works; in three of those four forces, the effects of ice can result in very nonlinear, thus non-proportional, outputs (we can ignore the weight of ice unless we are flying a Zeppelin). For example, when the ailerons on Simmons 4184 snatched and the yoke was literally pulled right

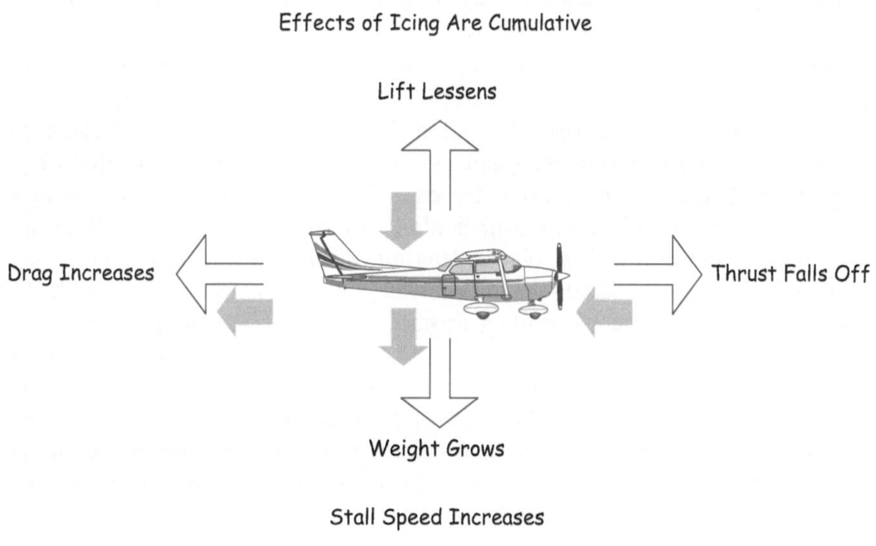

FIGURE 5.1 Historical representation of icing effects.

out of the first officer's hands, that was hardly a proportional response to the change between four and six degrees in AoA resulting from the flap retraction. Further, the loss of longitudinal stability experienced by Viscounts, Convairs, Jetstreams, YS-11s, and even DC-9s due to ice accreted on the horizontal stabilizer was never a proportional response to either an additional five or six thousandths of an inch of ice or to a couple of degrees de-cambering of the stabilizer with airplane-nose-down elevator movement. Or consider that the radically altered drag polars resulting from ice accretion yield a very disproportionate airspeed decay with increased AoA during the landing flare, resulting in countless hard landings with ice accreted.

And yet we cannot shake the idea of linearity. Besides the ubiquitous idea that the effects of ice are cumulative, we have an icing severity index, published in the AIM for decades and incorporated in several FARs, which clearly equates increasing thickness of ice with a linear degradation in performance.[13] The aerodynamic facts are that the effects of ice on lift, drag, and control surface balance may initially appear to be proportional to increasing ice thickness, but at some earlier-than-expected AoA, in concert with some seemingly arbitrary ice shape, they will become nonlinear and very non-proportional, often with catastrophic consequences. The insistence on a linear explanation and model of icing effects has probably cost quite a number of lives.

Because on the one hand, linearity is a mathematical concept, but on the other hand, it is important to us in the sense that it represents the intuitive relationship between cause and effect. In particular, it represents a single effect resulting from a single cause. The reason Laplace thought a vast enough intellect could eliminate uncertainty is because he thought that, with enough data, you could get that straight line on the X-Y plot just about right, at just the optimal slope, so that you would be assured of knowing a singular, unique value of X for every single value of Y you plugged in. If you know the cause, you can predict the effect; if you know the effect, you can backtrack to the cause.

And if, as Bruner has suggested, we are trapped within two possible modes of thought, we should consider the role that linearity plays in the narrative mode of thought, since that is the mode of thought that dominates the understanding of specific risk and that almost always emerges from an accident investigation report. Bruner's narrative skeleton concludes with a coda or moral to the story.[14] The coda will essentially be some form of a statement saying "if you want to avoid the effect, the peripeteia, then you must avoid the cause". Bruner's narrative skeleton may become dysfunctional when faced with nonlinearity, because the cause of the peripeteia may be undetermined or, if it is determined, inexplicably non-proportional, and thus the coda is rendered meaningless. The linear description of icing effects taught for so many decades is a likely artifact of the use of the narrative mode of thought, quite possibly left over from the era of the tenth axiom of the CAA Flight Instructor's Manual. And yet, the almost inevitable conclusion drawn from icing accident reports, even today, will be that the pilot allowed too much ice to accrete on the airplane. These kinds of conclusions emerge from an insistence on backfilling the narrative to arrive at a comprehensible coda, regardless of the remarkably nonlinear aerodynamics of an ice-contaminated wing.

The ADIRU spikes also exhibited a nonlinear dynamic. The investigation leaned toward a problem within a hardware component, "the CPU module, including the

CPU chip, application-specific integrated circuit (ASIC), a wait-state random access memory (RAM) chip, and four RAM chips for general CPU use".[15] The data spikes appear to have been random; however, if we dig deeper, this opens some interesting possibilities. The function of the ADIRU may contain elements of a complex physical system, and its seemingly random behavior may be chaotically deterministic.

CHAOS

In the late 19th century, yet another French mathematician, Henri Poincare, opened the door to 20th-century thinking about the ideas of chaos theory by offering the following observation:

> If we knew exactly the laws of nature and the situation of the universe at the initial moment, we could predict exactly the situation of that same universe at a succeeding moment. But even if it were the case that the natural laws had no longer any secret for us, we could still only know the initial situation *approximately*. If that enabled us to predict the succeeding situation with *the same approximation*, that is all we require, and we should say that the phenomenon had been predicted, that it is governed by laws. But it is not always so; it may happen that small differences in the initial conditions produce very great ones in the final phenomena. A small error in the former will produce an enormous error in the latter. Prediction becomes impossible ...[16] (italics added)

Stop for a moment and recall the forward process described by Taleb, the prediction of which he showed would flame out after a few iterations. He explained that as you consider points in time farther and farther into the future, you will need increasingly greater precision to model the dynamics of any particular process, because the error rate compounds rapidly. Eventually, you will have to achieve infinite precision.[17]

When Poincare says that "we could still only know the initial situation *approximately*. If that enabled us to predict the succeeding situation with *the same approximation*, that is all we require, and we should say that the phenomenon had been predicted, that it is governed by laws",[18] he is accommodating Taleb's point regarding the degradation of precision, and he is talking strictly about a linear system. But what he says next, "... it may happen that small differences in the initial conditions produce very great ones in the final phenomena",[19] is a fundamental shift in causal thinking, destroying the clockwork universe and departing linearity, leaving Laplace in the dust. Here, he is referring specifically to nonlinearity. Poincare was on the cusp of discovering chaos.

The disruption of Newtonian scientism and the Clockwork Universe is probably best described by the physicist Michael Berry and his example of the problem of billiard balls. In a 1988 interview, Berry explained it this way:

> We, as theoretical physicists, were brought up to believe that ... before the advent of quantum mechanics the universe was predictable in the sense that you had, for example, Newton's laws of motion, which, even as modified by Einstein, tell you that if you know the initial state of the universe – all the particles and their positions and velocities – you can predict for ever more its behavior. We believed that. We swallowed that particular myth although it flies flagrantly in the face of anybody who has ever

> used a pinball machine. But in fact, what's realized now is that unpredictability is very common, it's not just some special case. It's very common for dynamical systems to exhibit extreme unpredictability, in the sense that you can have perfectly definite equations, but the solutions can be unpredictable to a degree that makes it quite unreasonable to use the formal causality built into the equations as the basis for any intelligent philosophy of prediction ….
>
> … the more realistic case, actually, is if you think of the particles as billiard balls on a billiard table – an ideal, perfectly flat, perfectly smooth table – and this time you're considering the uncertainties as being produced by the gravitational force of people moving about near the billiard table. There always are such people milling about. You want to know after how many collisions of the billiard balls will their motion be uncertain because of this. And the answer is six or seven, and that's why no billiard player, even the best in the world, can ever plan a shot which would have even three or four consecutive collisions and have some reasonable expectation that he'll be able to successfully carry it out.[20]

Consider a problem that appears, at the outset, to be as simple as the billiard ball problem: the classic, first-year physics problem of the pendulum. The motion of a simple pendulum can be solved easily only within a narrow scope; if you let it swing beyond a few degrees, the simplifications used for a small period of oscillation disappear and the math gets pretty tough. But even that math pales in comparison to the problem of a *double* pendulum. In that case, calculus is quite inadequate to describe the pendulum's motion. The behavior exhibits an apparently erratic pattern of swings, flips, and reversals. It is chaotic.

Figure 5.2 shows what this might look like. The initial arc of the pendulum looks pretty much like what we would expect a simple pendulum to look like. Very quickly,

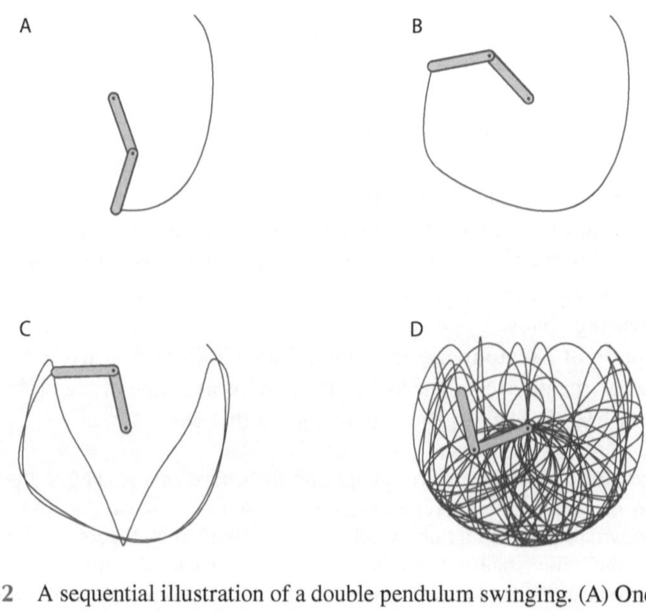

FIGURE 5.2 A sequential illustration of a double pendulum swinging. (A) One-half swing. (B) One swing. (C) Four swings. (D) Multiple swings.

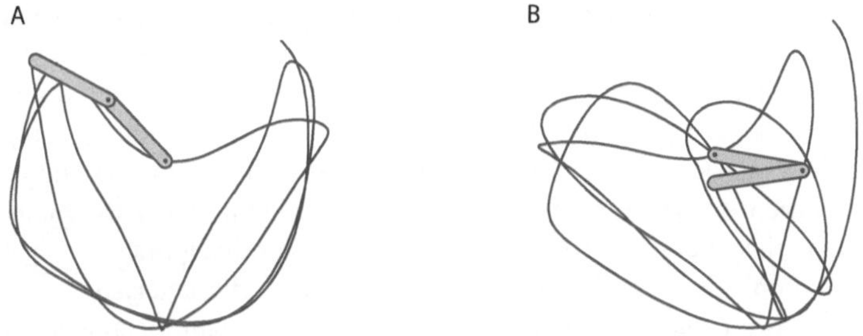

FIGURE 5.3 The sensitivity of a chaotic system to initial conditions. (A) Ten seconds, release angle = Thirty degrees. (B) Ten seconds, release angle = Twenty-nine degrees.

though, the double pendulum begins to exhibit some seemingly random behavior; if you let it swing long enough, you get a very chaotic-looking pattern. That doesn't look like anything we've seen in school.

Chaos theory introduces a concept that is critical to our understanding of prediction and iteration: a chaotic system is incredibly sensitive to initial conditions. The very slightest alteration in the pendulum's release point yields a radically different pattern of swings and flips. From whence we get the concept of the butterfly effect, the idea that the flap of a butterfly's wings somewhere in the world can lead to a hurricane somewhere else in the world. Figure 5.3 shows how sensitive our double pendulum is to a one-degree change in the release angle.

If you documented enough sequences of the double pendulum motion, you may find that the patterns tended to approximate, but never quite duplicate, one or two general patterns of swings and flips. This, too, is enormously dependent on the initial conditions, including such things as friction, air resistance, and gravity as well as the precise mass and dimension of the pendulum structure. In chaos theory, these patterns are known as strange attractors. If they appear, the probability that the next swing will be close to one of these general patterns is fairly high. Moreover, if you knew exactly which swing or flip the pendulum had just completed and could measure it with perfect precision, you could predict the next swing or flip. The problem is that your ability to accurately measure the preceding swing is limited by the imprecision cited by Taleb and Berry. And the slightest error … one decimal place several million decimal places out … can yield a completely different behavior.

COMPLEXITY

Nonlinearity and chaos theory are stepping stones into the vast world of complexity. And complexity has the potential to completely change our understanding of system behaviors, failures, and accidents in general. The challenge is that it cannot be described using the narrative mode of thought. It is, at best, paradigmatic, and perhaps not even describable by that. This leads to a wide range of definitions, even from people who have pioneered these concepts. Further, one of the problems with integrating and synchronizing complexity into our worldview is that it is applicable to a very wide and diverse range of systems and disciplines, which runs quite counter to contemporary, enlightenment concepts of expertise and administrative organization. An early challenge in the development of this concept took place at the very first conference held by the Santa Fe Institute, which had been chartered for the specific purpose of studying complexity. That first meeting was primarily comprised of economists and physicists. The initial challenge was to find common language with which they could discuss their viewpoints with each other and then establish the idea that the mechanics of complexity were universal to their disciplines.

As such, complexity is probably best described by its characteristics.

Complexity is a behavior that is rich, collective, and dynamic, created from simple interactions between very large numbers of components. This introduces a fundamental aspect to complexity and complex systems; they are comprised of a very large number of often very simple components.

John Holland, of the University of Michigan, is one of the earliest pioneers of complexity. In his book "Complexity: A Very Short Introduction", he has identified four major characteristics of complexity:

- Self-organization into patterns, as occurs with flocks of birds or schools of fish.
- Chaotic behavior where small changes in initial conditions produce large later changes.
- "Fat-tailed" behavior, where rare events (e.g., mass extinctions and market crashes) occur much more often than would be predicted by a normal (bell-curve) distribution.
- *Adaptive interaction*, where interacting agents (as in markets or the Prisoner's Dilemma) modify their strategies in diverse ways as experience accumulates.[21]

Melanie Mitchell, in her book "Complexity: A Guided Tour" has proposed this definition:

> A system in which large networks of components with no central control and simple rules of operation give rise to complex collective behavior, sophisticated information processing, and adaptation via learning or evolution.[22]

At this juncture, it is important to differentiate between the words "complicated" and "complex". A complicated system is also comprised of a large number of components, but its output is linear. The superposition principle applies. A complex system is nonlinear, but more importantly it exhibits three characteristics which are distinct and counterintuitive. The first is known as self-organization, as described by Holland and also by Mitchell when she talks about "complex collective behavior". The second is emergence. The third is that of an open system which modifies through feedback. Let's take these ideas one at a time.

Self-Organization

In keeping with our adherence to fairly simple examples like billiard balls and double pendulums, a somewhat intuitive example of self-organization is a sand pile. In the late 1980s, a physicist named Per Bak developed a computerized model of the

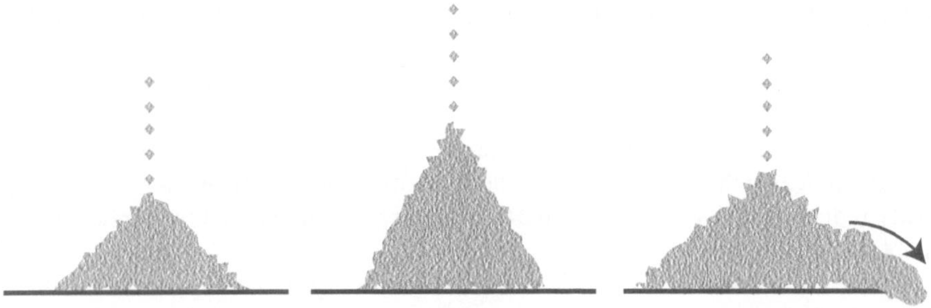

FIGURE 5.4 Per Bak's self-organized criticality.

behavior of a sand pile, shown in Figure 5.4. The actual math isn't important for our purposes. But consider a conical sand pile and think about what happens if you pour a steady stream of sand grains, one at a time at a constant rate, directly on the apex of the conical shape. Eventually you will reach a point which Bak called criticality; beyond that point, you have no way of knowing what will happen when the next grain of sand lands upon the apex. Ultimately, there will be an avalanche. It may be a small one or it may be a large one. The outcome will be a wider base, a shallower slope, and a return to stability well below criticality. Until the process repeats with the addition of more sand.[23]

Bak referred to the behavior of the sand pile model as "self-organized criticality". The avalanche is a rich, collective, and dynamic behavior that results from the simple interactions between large numbers of sand grains. Exactly when the avalanche will occur is indeterminate, as is the size of the avalanche and the final point of stability. All of these factors result from the physical interactions between sand grains … the mechanical forces that dictate the motion and stability of each sand grain with respect to its immediate neighbors. The phenomenon is "self-organizing" because while it results in an organized, symmetrical system of sand grains, it does so without any sort of future purpose or by following any type of master blueprint. No sand grain is in possession of any encoded information other than the loads applied to it by the sand grains immediately adjacent. Clearly, no sand grain has any concept of the shape of the sand pile nor any particular hopes or plans for what that pile will look like after the next avalanche. Indeed, no sand grain has any idea what a sand grain on the other side of the pile is experiencing, or that there even *is* a sand grain on the other side of the pile. While this may sound a bit tongue-in-cheek, the point is that the sand pile will reorganize itself without a central authority or controlling component. This is an incredibly important aspect of complexity.

Consider a further analogy: a symphony orchestra. This is the kind of thing we normally envision when we think of a vast number of parts working in a system. However, the sound produced by a symphony orchestra is actually linear, in that the superposition principle applies. The output is a summation of the inputs; flute plus saxophone plus violin plus drum equals the music. If you add together the sound made by the flute plus the drum, and then add the sound made by the saxophone and the violin, you will get the same sound as you would have if you had started with the flute and the violin and then added the drum and the saxophone. While the output of the symphony orchestra may sound to us like a "rich, collective dynamical behavior", from the sound technician's point of view it is quite easily reducible to its component parts. A symphony orchestra is complicated; it is not complex.

Further, it has two characteristics which a complex system does not have. Unlike Bak's sand pile example, the symphony orchestra has one component who is in charge: the conductor. A complex system has no conductor. Second, in the orchestra, each component … musician … has the capacity to understand the function of the entire orchestra. Such is not the case in a complex system: were the orchestra actually complex, the first violin wouldn't have a clue what a cello is.

The classic example of self-organization within a complex system is an ant colony. Since the individual ant is a living creature, quite possibly in possession of some element of free will, it is again easy to ascribe organized behavior to a pre-organized master plan, perhaps instinctively encoded. In many ant species, there is only one queen ant to the colony. Because of the larger size of the queen and more likely because of the meaning we take from the word "queen", it is easy to envision the queen ant as being the conductor of the orchestra, the royal sovereign in charge of the whole shebang. Of course, this is not true … the queen is the mother of all ants, but certainly no commander-in-chief. This leads to the obvious question in the narrative scheme of things: who is the commander-in-chief?

Well, once we have abandoned the idea of a royal sovereign, our next default position is to the soldier ants, those muscular tough guys who defend the colony. Since, in our normal conception of things, the soldier ants appear to be exhibiting virtuous and responsible behavior, demonstrating loyalty and correctly prioritizing everything in their little Hobbesian world, we can imagine that they have the big picture, an overall leadership understanding of what has happened, is happening, and will happen, and what must be done next.

An ant's brain contains about a quarter million neurons. The human brain is thought to contain between 80 and 90 *billion* neurons. We encounter humans every day who seem to have difficulty grasping the big picture; it seems unlikely that any particular ant may be able to grasp even a small picture, let alone the whole panorama. So, without a royal sovereign, without any responsible adults under our conventional definition of that term … how do they get anything done? They self-organize into patterns of behavior. This is not a directed response; it is the result of local interactions based on local rules. It is known that individual ants communicate through chemical perceptions experienced by touching antennae. It is believed possible that ants maintain a sort of statistical tally of how many individuals are executing which tasks and use this local knowledge to fill in and balance the workload.

The notion that there is no single component in charge flies in the face of practically every organizational concept known to man. It is an enormously difficult idea to wrap one's head around. The descriptive terminology used for a CPU itself promotes the notion of hierarchy; what is a CPU but a "Central" Processing Unit, the brain of the computer? And indeed, within the CPU is a "control" unit, which extracts and executes coded instructions. Terms such a "central" and "control" are consistent with an image of hierarchy, a top-down management of the system. They very strongly imply that there is someone, some component, in charge of the organization.

But neither the CPU nor the control unit within the CPU is capable of comprehending or envisioning the overarching function of the computer or even possessing the slightest notion of purpose. They really are not "in charge" of the computer or the CPU. While the software for extracting and executing coded instructions is a "top-down" process, that process can be heavily influenced by the "bottom-up" physical interactions between the components of the CPU.

Holland has defined two types of complex systems: a complex physical system and a complex adaptive system.[24] The fundamental difference is the ability of the system to adapt and evolve, which is the difference between explaining a sand pile or a CPU fault and the rebuilding of a flooded ant colony or, quite possibly, the learning process within the human brain. Both Holland's point regarding adaptive interaction and Mitchell's point about adaptation via learning or evolution apply to complex adaptive systems. For the moment, we will stay with the complex physical system, because that is what might have been operative with the ADIRU failure on Qantas 72.

Emergence

The second principal characteristic of a complex system is emergence. This is also a very challenging idea to grapple with. The simplest example of emergence is wetness. There is absolutely nothing in the characteristics or qualities of hydrogen atoms and oxygen atoms to suggest the quality of wetness, yet when properly combined into water, the characteristic of wetness emerges. In the paper, "Types and Forms of Emergence", Jochen Fromm, writing in collaboration with Russell Abbott, made this key observation: "The surprising aspect in a process of emergence is the observation of an effect without an apparent cause".[25]

Fromm has described emergence as a paradox; emergent characteristics are "changeless and changing, constant and fluctuating, persistent and shifting, inevitable and unpredictable". He goes on to say that

> … because true emergent properties are irreducible, they cannot be destroyed or decomposed – they appear or disappear instead. In this sense they may seem to be indestructible and are potentially the only things that really exist, but if they are examined too closely – if we take a deeper look at the components of the system – they do not exist at all and often vanish into nothing. The paradox arises mainly because we are often only able to see part of a complex system, if we consider only the microscopic *or* the macroscopic level, but not both at once, or if we see only the system *or* the environment, but not both. It is difficult to comprehend a complex system with multiple levels and scales.[26]

Fromm goes on to suggest a careful and meticulous taxonomy for emergence, but he relies completely on the paradigmatic mode of thought, using logical relationships to build his taxonomy. In those descriptions, he is able to show a gradient of predictability. His taxonomy shows that, at the weakest level, emergence is predictable because it is repeatable; while the property of wetness is not reducible to the component properties of hydrogen and oxygen, wetness is indeed repeatable, and so is predictable. Aside from the weakest level of emergence, however, he shows that precise resolution of emergent behavior is impossible. This becomes a powerful problem to overcome when attempting to construct a safe operating space within a complex system because within the traditional narrative story, the final element in Bruner's narrative skeleton, that of a coda or moral of the story, is apparently disconnected from the rest of the narrative. If the coda describes a true emergent property, then, using Fromm's words, it is irreducible and cannot be destroyed or decomposed – it appears or disappears instead. Without being able to link the coda to the rest of the narrative, the story has no value, because the effect of emulating the narrative cannot be predicted.

Emergence then appears at higher levels in the system than the level at which the constituent components reside. At the atomic level, there is no indication whatsoever of the property of wetness; however, at a higher level, wetness appears without any apparent cause. This definition appears in the QF72 report, when the investigators said that "A much more likely scenario is that the [data-spike] failure initiated a problem in a higher-order process for organizing the storage or retrieval of buffered data within the CPU module".[27] Fromm has also, coincidentally, pointed out that "Unforeseeable failures and unexpected faults in software and hardware systems are a special, undesirable form of emergence".[28] No doubt the crew of QF72 would consider this an understatement.

But we have to be careful in how we express this idea. Note that the Australian investigators described the relationship between the data-spikes and the "higher-order process" in a linear, cause-and-effect narrative, following Bruner's model precisely. They cited the data spikes as the *cause* of a problem in a higher-order process. This is analogous to stating that the molecular assembly of two hydrogen atoms and one oxygen atom is the *cause* of wetness. Buried within this description are two distinct levels of emergence. The first occurs at the physical level of the CPU itself; within that structure, the data spikes are emergent. The soft fault demonstrated that the spikes were not necessarily repeatable, and yet they did repeat numerous times during the event. No cause was ever identified, very probably because we do not know enough about the relationships and interactions within the CPU, at the molecular or atomic levels, to identify a cause. In this sense, the data spikes appear as properties that "cannot be destroyed or decomposed – they appear or disappear instead".[29] Once they do occur, a second level of emergence occurs within the process for organizing the storage and retrieval of buffered data; this process may be weakly emergent, in that it is predictable in principle but not in any specificity. Knowledge of the data spikes alone would not have allowed the specific prediction of the buffered data failures.

Emergence is the destruction of Newtonian linearity. It is with this understanding that we can begin to appreciate the futility of the tenth axiom, the notion that a capable and competent aviator will never allow a crack-up. It is contributory to understanding why you will never see the one coming that gets you. The key to emergence is irreducibility; it is greater, or lesser, than the sum of the constituent components. You cannot simply account for all of the properties of the constituent components and expect to get the emergent property.

Open Systems

This brings us to a third important characteristic of complex systems: they are open systems. The scientism that is accessible to most people is predicated on the revolution in scientific thinking expressed through Newton's work in the 17th century. It is based on universally applicable physical laws. For example, Newton quite accurately said that for every action, there is an equal and opposite reaction. That this is true is undeniable; Mike Collins would certainly have attested to it after the effort he expended while space-walking to retrieve the cosmic dust collector from the second, abandoned Agena during the Gemini 10 mission in 1966.[30] Indeed, Newton's laws, and others, were extremely effective in sending man to the moon and back, including the wizardry of slinging the Apollo 13 crew around the moon, and leveraging lunar gravity, to get them back to earth.

But the application of Newton's laws requires a very narrowly constrained reality which, generally being linear, coincidentally fits well into the narrative mode of thought. Newton defined a canonical state of things, and thus, by definition, a Newtonian system is a closed system. But the fundamental nature of the world is that no system is truly closed. All systems, when moved from the design room into the real world, become open … and this can have important implications for complexity.

Open systems are vulnerable to a myriad of local information. Sidney Dekker, in his book "Drifting into Failure", has explained open systems this way: they can exchange information and energy not only internally, between components of the system, but externally with the environment. The local interactions between adjacent components are generally minor but take place outside of any canonical order or general purpose. An external input of sufficient energy can modify the local rules at the point of contact with the system.[31]

A complex system is capable of responding to feedback. Feedback can be described as the output of a system being returned to the system as an input, which modifies the variables and results in a different output and consequently a different feedback. While the initial condition may be a local response to an external input, the chain of local interactions begins to energize in response not just to the external input, but also to adjacent internal inputs. This development of response can amplify through the system, leading to local responses at the other end of the system that take place, again, in the context of local rules, without any sense of the relationship to the original external input. The eventual output, if fed back into the system as an input, then results in a very different output because the local rules have been altered.

The Australian investigators were sufficiently intrigued by this possibility that they carefully investigated a phenomenon known as a single event effect, or SEE. This phenomenon involves an interaction with a high-energy particle generated by a cascade effect following the entry into the atmosphere of solar and or galactic radiation (cosmic rays in lay terms). The report stated that

> A typical computer silicon chip, or integrated circuit, is primarily made of a large number of tiny devices called transistors, each of which operates by accumulating and distributing electric charges within the chip. A subatomic particle that collides with an atomic nucleus inside a chip can change the amount of electric charge within a particular area (a phenomenon known as localized ionization). If sufficient charge accumulates in a particular area, a transistor's behavior can change.[32]

In the end, they were not able to establish any so-called trigger events for the ADIRU CPU behavior. However, while they did feel that they had eliminated several internal causalities, they clearly felt that they could not definitively rule out SEE.

(As an aside, back in the mid-eighties when the Airbus fly-by-wire technology first started to emerge, I frequently commented to ground school classes that I thought this was a really dumb idea; when students would point out that the latest iteration of modern jet transports did not have any manual reversion anyway, I would ask how many stray quarts of hydraulic fluid were wandering around the upper atmosphere, as opposed to how many stray electrons? It was vaguely tongue-in-cheek, and it turns out that this possibility had actually crossed the minds of the Airbus system designers. But SEE fits the description).

From the standpoint of analysis, an open system creates a real conundrum. When I was in college, one of the standard concepts we used in the study of thermodynamics, fluid dynamics, and so forth was that of the control volume. We arbitrarily enclosed the system so that we could control the measurement of energy in and energy out. It turns out that this technique is pretty much the only way we can think about physics and natural processes. It is a subset of the traditional techniques of modelling nonlinear systems with pieces of linear systems. We try to position our control volumes so that the inputs that we cannot measure, or over which we have no control, can be ignored, allowing us to analyze specific parameters of interest. Failing this, we construct analytic models which bypass problematic parameters. This is the utility of the Eulerian approach to analyzing fluid flow, by assuming zero friction between the fluid and the pipe, when in fact the system is very much subject to viscosity and the nonlinearity of Navier-Stokes analysis. The Eulerian approach yields a usable result, but not necessarily one that works within a system that is hyper-sensitive to initial conditions strongly influenced by external inputs. It also runs on far smaller computers.

What's more, if we are limited to analyzing an open system using finite structures of closed systems, we are even more hamstrung by the inability to predict the progress of consequence between those analytic model structures. In a finite system comprised of a limited set of components, we are able to ignore small perturbations to the system and accurately, for our purposes, predict the output at one end of the system based on the input at the other end. Such a prediction is processed through known relationships between components that allow a consequence of an input to be propagated through the system. We can predict this; in a clockwork, when the counterweight drops, the gears will all turn in a predictable fashion and yield a movement of the clock hands. The progress of consequence is easily mapped. So it was imagined for the clockwork universe envisioned by Laplace, and this is what Poincare was referring to when he said that "we could still only know the initial situation *approximately*. If that enabled us to predict the succeeding situation with *the same approximation*, that is all we require, and we should say that the phenomenon had been predicted, that it is governed by laws".[33]

In a complex system, the progress of consequence is nearly impossible to map. Each component responds locally to the information available in the moment. This information is not held in reserve for any future purpose, as there is no component aware of any future purpose. However, the passage of information between components can modify the rules binding those two components, such that on the next iteration, a different consequence is produced and the rules are again modified, but in a different manner. Because the rules and relationships that connect any two components can be modified by the passage of information through those relationships, a complex system has a history. The state of the system is defined by how the system has previously interacted up to that very moment. What it just did is the launching point for what it will do next.

With those characteristics of a complex system in mind, we can consider how they apply to commercial aviation safety. We really operate within three inherently complex systems. One system is the aircraft itself. Whether it be through phenomena such as arc-tracking, or the changing aerodynamics resulting from an ice accretion that itself changes because of the changed aerodynamics, or multiple data spikes in one of the ADIRUs, complex physical system behavior is possible within the aircraft as a system. This idea is new to us in the sense of automation. We didn't use to have a lot of automation, and what we had could hardly be defined as complex. Today we operate aircraft via huge numbers of continuous computations; there is an ever-present possibility of complexity making this go less smoothly than we thought it would. We can expect more types of emergent behaviors that appear, as Jochen Fromm has said, as an "observation of an effect without an apparent cause".[34]

Complexity in the Airplane

The phenomenon of arc tracking provides a useful example. Arc tracking has been known for many, many years; in the late 1980s, it became a focus of attention within the aviation safety community due to several incidents involving Kapton wiring.

In November 1999, Alaska Airlines Flight 400, a Boeing 737-400, experienced multiple fuel pump low pressure lights shortly after takeoff, accompanied by two fuel pump circuit breakers popping. The crew turned back to Portland, Oregon, and landed safely. Maintenance technicians subsequently located one hundred and thirteen wires that were damaged from electrical arcing.[35] On December 29, 2000, the crew of Delta Airlines Flight 219, a Lockheed L-1011, spent one hour and fifteen minutes fighting a recurring cockpit fire while enroute to Hawaii over the Pacific Ocean. Subsequent examination found twenty wires burned, some fused into solid copper over more than one half inch of length, others with re-solidified molten globules at the severed ends.[36] And in the spring of 2006, the NTSB investigated seven electrical fires in Canadair Regional Jet avionics compartments, at least two of which resulted in loss of EFIS displays and one of which resulted in the loss of generators, electrical smoke in the cabin, and loss of cabin temperature control.[37]

Arc tracking is created when an external conduction path is generated on the insulation surface of a wire. This can begin with a simple mechanical short circuit, due to abrasions or cuts in wire insulation, or it can be initiated by the degradation of wire insulation. An electrical arc is a non-linear, chaotic event. As insulation degrades, micro-arcs can occur between the conductor inside the insulation and the external atmosphere. These micro-arcs are very weak but sufficiently hot to create a charred spot on the surface of the insulation, known as pyrolization, in which the insulating material is transformed into carbon. Carbon is a conductor. Eventually, the proximity of carbonized spots leads to an arc which jumps the spots, charring a length of insulation. This opens up the possibility of further arcing, and eventually a cascading, "avalanche" arcing event occurs causing ignition of the wiring harness and any surrounding flammables, such as dust or debris. Figure 5.5 shows an example of the result. This behavior is not unlike Per Bak's sandpile avalanche.

Using Jochen Fromm's taxonomy of emergence, arc tracking is a form of "weak" emergence. This can roughly be defined as a characteristic that is "predictable in principal, but not in every detail". Fromm goes on to explain that

> In weak emergence, there is no unique direction of causality from the microscopic to the macroscopic level as in simple forms of nominal or intentional emergence. There are causal relations in both directions.[38]

The dynamics of arcing create a feedback loop in the system. Once the arc takes place, the pyrolization of the insulation creates a foundation for additional arcing. The more arcing takes place, the more pyrolization, until it is possible for greater and greater amounts of energy to be released through a larger arc, up to the power capability of the system. While the initial micro-arcing is a function of local rules … condition of the insulation, moisture presence, surrounding electrical charge … the flashover that finally occurs is a function of highly modified rules operating in an open system, applying energy externally to adjacent, intact wiring, initiating rapid pyrolization, and combustion of that material.

It is probably safe to say that micro-arcing, and the resultant pyrolization of tiny, microscopic spots on insulation material, has taken place in many, many wire bundles that use older insulation compositions like Kapton. Clearly, not all of these events result in arc tracking and flashover. The problems identified in the CRJ avionics bay almost assuredly existed in the entire fleet of CRJs that were equipped with 1K4XD contactors using Ultem 2200 as a base material, as opposed to those using the originally certificated NP509 woven glass fabric, melamine resin laminate material, which did not result in any electrical anomalies. Thus, the arc tracing characteristic in this case is "predictable in principal, but not in every detail".

Complexity in the Atmosphere

The second complex system is the atmosphere, the complexity of which becomes ever more apparent as we gain measurement capability and data. Indeed, the modern understanding of chaos theory and in particular, the sensitivity to initial conditions, began with the Massachusetts Institute of Technology mathematician and meteorologist

NASA
C-1996-1430

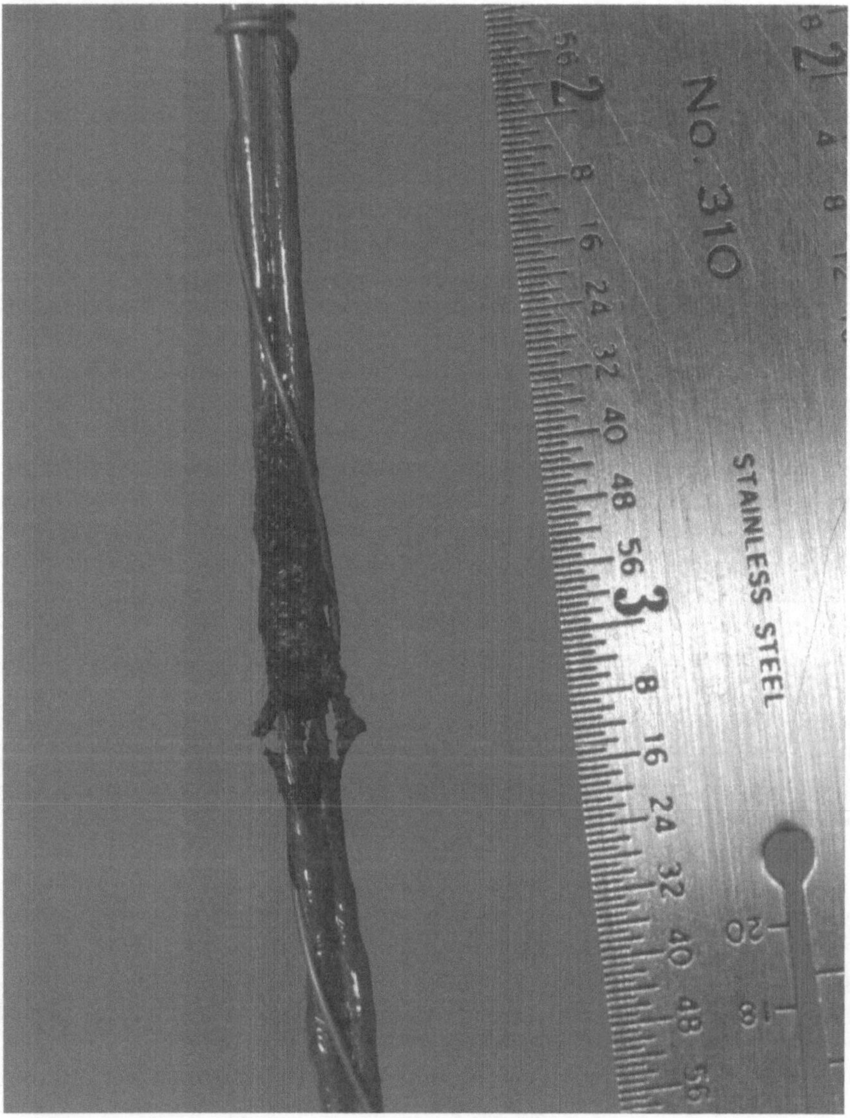

FIGURE 5.5 Kapton wiring damage due to arc tracking.

(From National Aeronautics and Space Administration Lewis Research Center.)

Edward Lorenz, who in the early 1960s was attempting to use a computer model to simulate weather prediction. In the process of re-running a particular set of data, he chose to shorten the process by initiating the model mid-way through an earlier run, which he did by entering the inputs captured at that point of the model run on a printout of the earlier iteration of the same data run. Much to his surprise, he got a completely different result. He had stumbled upon a sensitivity to initial conditions; the data he used from the printout had been truncated and rounded several decimal places into the values, but that difference was enough to yield a radically different prediction.[39]

Within the field of aircraft icing, we also encounter clear cases of weak emergence. Icing is really a prime example of Fromm's description of emergence as "changeless and changing, constant and fluctuating, persistent and shifting, inevitable and unpredictable".[40] We can see this in the wide range of pilot reports that are often extracted from a particular icing environment, ranging from light rime to severe clear. These reports are often not only dependent on aircraft scale but also on timing and aerodynamic properties such as AoA and airspeed. As discussed earlier, we often attempt to linearize this phenomenon, using such phrases as "that airplane can carry a lot of ice" right up through the AIM definition of moderate ice, "The rate of accumulation is such that even short encounters become potentially hazardous and use of deicing/anti-icing equipment or flight diversion is necessary".[41] Yet the actual presentation of inflight icing is very much dependent on the state of the system immediately before. The icing environment is easily characterized as predictable in principal, but not in every detail. So, too, is the shape of the accretion of ice, and lastly, the aerodynamic effects of ice accretion are certainly predictable in principal, but not in every detail.

Fromm has divided weak emergence into two families, that of stable emergence and unstable emergence.[42] Stability results from negative feedback, a type of feedback which dampens the emergent development. Instability results from positive feedback, in which the feedback amplifies the emergent development. The arc-tracking events that we do not know about are essentially stable; the feedback generated by pyrolization is inadequate to accelerate the arcing process. The cases we do know about are unstable, which is why we know about them. The pyrolization is more than adequate to accelerate further arcing and pyrolization, until the avalanche.

Whether the icing situation is a stable form of weak emergence or an unstable one must be considered carefully.

An icing environment is stable, in that there is a process of negative feedback involved. A parcel of air can only accommodate a limited amount of water; while it can be quite a lot of water, resulting in severe icing, the physics involved limit the unbounded expansion of the icing environment. Statistically, it is a bounded environment; there aren't going to be any fat tails, although the distribution is rather more complicated than a simple bell curve. The available liquid water will be distributed in droplets of various sizes, themselves dependent on a range of inputs … local rules. The water may appear in large droplets, which will inevitably be associated with low liquid water content in a given parcel of air. Or they may appear in small droplets, associated with greater liquid water content. There is a stability here; there is an upper end to droplet size and liquid water content. Eventually, often rather quickly, the cloud will glaciate. This is a briefly unstable process, known as the Bergeron process, in which water vapor will sublimate on the ice crystals. This reduces the vapor pressure and draws liquid water from the existing droplets into vapor, which then also sublimates on the ice crystals, until there is no more liquid water. The process of glaciation is highly dependent on initial conditions, that is, the number of cloud condensation nuclei and/or the presence of water droplets that freeze homogeneously at temperatures below –40°C. It is an unstable, accelerating process. Obviously, once the supply of liquid water runs out, the process stops and the cloud is considered fully glaciated.

On the other hand, the aerodynamic accretion of icing may be quite unstable for an extended period of time. Instability is the result of positive feedback; the more ice is accreted, the greater the surface area available for more ice accretion. Eventually, stability in the accretion process is regained. There are limits on the size and shape of an ice accretion in a given aerodynamic environment. But for a critical period during the early accretion of ice, the process may experience positive feedback and be unstable. Moreover, the development of an ice shape is an emergent property, in that it is the result of many local interactions between droplet size, liquid water content, temperature, freezing fraction, airspeed, and the scale and shape of the accreting surface. While the ice shape may be nominally reducible using computational fluid dynamics, this becomes quite problematic in the face of external inputs to an open system. Minute differences in the initial conditions can lead to significant differences in the ice shape itself and its characteristics, for example, roughness.

Figure 5.6 shows a typical double-horned ice shape developed on a test airfoil in the NASA Icing Tunnel. As with the pattern of swings in the double pendulum, the limits of the shape are predictable; the actual physical character of the ice is not. It is emergent.

The aerodynamic effects of ice accretion exhibit an inherent instability, insofar as the increase in drag provides positive feedback to the decrease in airspeed,

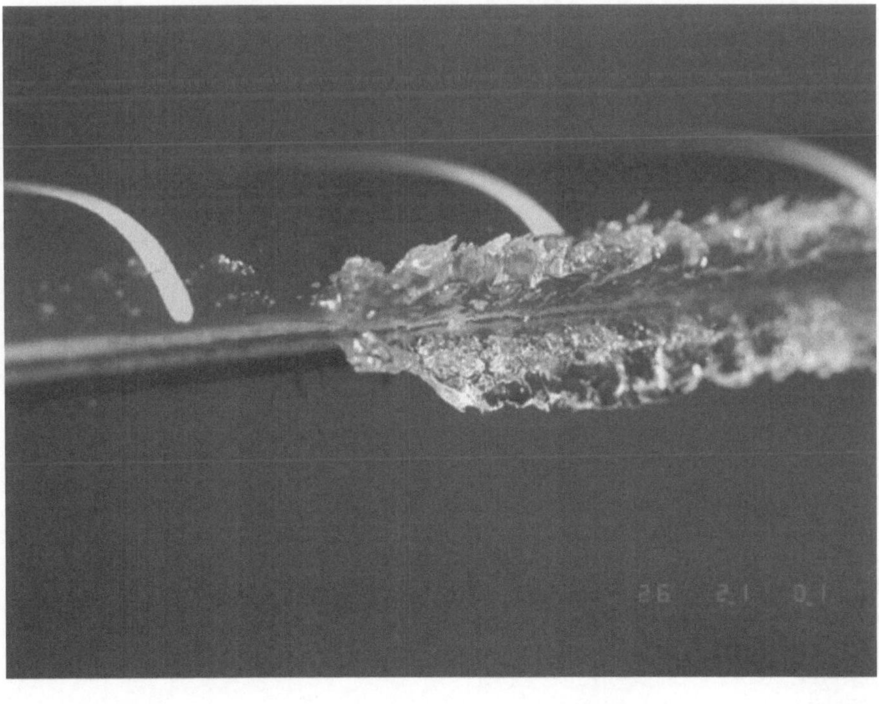

FIGURE 5.6 Ice accretion on a test airfoil in the icing wind tunnel.

which in turn leads to an increase in AoA, providing further positive feedback to the aerodynamic situation and to the ice shape itself. There is no negative feedback in this process to slow or otherwise inhibit this effect. How rapidly this occurs is also dependent upon the initial conditions, that is, the state of the system immediately preceding. There will eventually be a radical change in the aerodynamics … a stall, or a control surface snatch … a sort of cliff, beyond which nonlinearity dominates, but when and how this occurs is entirely dependent on the ice shape and the state of the aerodynamics immediately preceding the cliff.

Thus, the icing problem is emergent. The icing environment is emergent, the ice shape is emergent, and the icing aerodynamics, particularly following flow separation, are emergent. We consistently make a fatal mistake when we interpret, often through the normalization of deviance, the icing problem within a linear, closed system model.

And that's not the only emergent phenomena with which we are already familiar in the atmosphere. One of my favorite forecast notes, which appeared one morning a few years ago in the Dallas-Fort Worth forecast discussion, simply stated, "To convect or not to convect; that is the question". A thunderstorm is a spectacular example of Fromm's concept of weak emergence … predictable in principal, but not in every detail. As Figure 5.7 shows, thunderstorms all look pretty much the same, but no two are identical. Another forecast for the Chicago area captures the problem well:

> Today provides quite the complex forecast for thunderstorm evolution and timing. Even just by later this morning, there is a large spectrum of possibilities. For a bottom line

FIGURE 5.7 A thunderstorm as a complex system exhibiting weak emergence.

> up front, there is the likelihood of some influence on evening convection by late morning showers and possibly some storms that move into far northern Illinois. How much influence, and quite possibly very little besides where outflow and/or the composite front sets up, is tough to determine at this juncture. There remains the likelihood of a general west-to-east axis of convection, possibly two though one being more dominant, from Iowa into northern Illinois by early-mid evening. This would likely continue for several, maybe even 6+ hours, with a scattered severe threat and a potential for a stripe of flash flooding concern. While within a broad Slight Risk that speaks well for the pattern and uncertainty, the scattered severe threat may be more locally realized in north central Illinois than the rest of the CWA, but again that is based on some influence of late morning convection.[43]

Within this forecast, we can see a general theme: what happens later today in Chicago is entirely dependent on what happens earlier today – the history which drives the emergence found within a complex system. Until we know that history, we have no good way of narrowing the prognosis. We can also see clearly that the system is open and will be strongly influenced by external inputs. These are important points, because thunderstorms represent another potential misapplication of linearity. We know a lot about them. We understand how they form, how they die, and we can differentiate between mesoscale systems, supercells, bowed echoes, and hooks. Any reasonable textbook discussion of a thunderstorm will describe their behavior and structure in detail. But conversely, a thunderstorm is clearly self-organized. It has no central control; the air and water molecules involved all self-organize according to local rules involving solar heating, water vapor, wind shear, release of latent heat, and many others, making the precise development of the storm inherently difficult to predict. The difficulty lies, as always with a complex system, in the sensitivity to initial conditions and in the non-linearity of the physical processes involved. As with most physical systems, we are constrained to study thunderstorms within a closed system. Reflecting the above Chicago forecast, this limitation results in statements such as the following from a research paper by Rebecca Cintineo and David Stensrud, "On the Predictability of Supercell Thunderstorm Evolution". They were explaining one of many limitations to their simulations of supercell development:

> One limitation of this study and its applicability to the real-world environment is that the initial environment is initialized with a single point sounding and is therefore horizontally homogeneous. The real atmosphere contains inhomogeneities, such as airmass boundaries and influences from topographic features, and other storms may already be within the environment at the storm's initialization time that could affect the storm's subsequent evolution through their outflow boundaries or interaction between the storms themselves.[44]

Finally, there is perhaps no better example of Per Bak's sandpile experiment in the aviation world than the microburst. The water loading in the upper part of the storm reaches self-organized criticality, fueled by the intensity of the updraft, followed by the complex self-organization of the downburst involving intense evaporative cooling, hail melting, dry air entrainment, and the weakly emergent behavior of straight line winds pressurized by the cold pool's local mesohigh. There is no good way to know exactly when the microburst will initiate or precisely how intense it will be;

only that, if it occurs, it will occur in a furious avalanche of self-organization, just like the sandpile.

Complexity in People

The third aspect of complexity in aviation is human behavior. At this point, we depart from John Holland's notion of a complex physical system and move to his idea of a complex adaptive system.

With many schools of management still stuck on Simon's early idea that two people will do exactly the same thing if given precisely the same information, we still act surprised when two people do something entirely different. Yet there are those 80–90 billion neurons, all interconnected through local relationships and local rules, all engaged in various degrees of feedback, all without any vestige of hierarchical control, self-organizing brilliantly while the human being that they breathe life into spends decades looking for inner ids, subconscious souls, hippocampi, and all manner of other structures capable of retaining memory for future purpose while coordinating that purpose throughout the entire body. Even more frustrating to the management class is the pernicious tendency of groups of human beings to work together outside of the official protocol, organizational structure, or other controls through which the formal hierarchy executes present and future endeavors.

I am always amazed at how the airline manages to realign a situation that has, for the moment, gone off the rails. When we take a fully loaded airplane out of service for, say, a mechanical failure that is clearly going to require hours of work to fix, we confront a host of problems. We have to deplane the passengers, remove the baggage and cargo, and transport all of this to a different airplane at a different gate. We have to refile the flight plan, recalculate the load plan, and transfer the catering from the broken ship to the new one. Or, in another example, when the weather necessitates a re-routing, we have to alter the flight plan, recalculate the fuel load, re-evaluate the alternates, and re-load the flight plan into the FMC. Because these activities are organized through a hierarchical set of procedures, the expectation is that the outcome is predictable, and, largely, it is. However, these situations are also perturbations in the normal flow of things, placing heavy burdens on time and resources. How they are reintegrated back into the routine, the prescribed organization set forth by regulation and management, is quite probably influenced by local rules. Different airport authorities, different subcontractors, different union locals, and a wide variance in physical infrastructure require the human agents involved … possibly dozens of them … to take basic level actions that are influenced by local feedback. Decisions and choices are made due to weather, the distance between the gates, the requirements of local airport authorities, and so forth, which, in theory, are checked against standard operating procedures to ensure that the centerline of the safe operating space is accurately tracked. And yet, the time and resource pressures create a feedback that can lead to individual cases of violation behavior.

We could argue the airport operation to be a form of ant colony, in which the individual ants have no idea of the bigger picture created through self-organization. But there is a fundamental difference; the airport is comprised of human beings, who differ from the components in nearly all other adaptive systems, in that each individual believes that he or she *does* have the big picture. Indeed, we are trained that way by corporate gurus attempting to paint a big picture of corporate purpose toward which each individual is working. The difference between the perceived big picture and the actual one opens the door to some pretty unpredictable feedback loops. The multitude of perceived "big pictures" can lead to unique forms of self-organization, themselves very sensitive to initial conditions, such as time of day, proximity to shift change, particularly challenging ambient temperature, and many others up to and including an individual's state of mind at that moment.

In conventional parlance, this is often referred to as a causal chain of events. The weakness in this approach is the assumption of linearity, the idea that the "chain" is made up of interconnected "links" which in fact create a picture of an actual, literal line in our minds. The chain is hooked to the situation at one end and tugged on from the other end, giving it not only linearity in structure but in time and motion. Likewise, Reason's cheese is highlighted by a vector depicting the trajectory of time, in which the accident manages to linearly pass through several slices of holed Emmenthaler.

But the chain of events is not linear, rather it is a structural organization of components in space and time. The same is true for the slices of Reason's cheese. It is not the linearity that we need to focus on nor the idea that breaking a single link in the chain destroys the linear trajectory. It is not the individual links or slices that cause the problem, but the structure that they form in that particular moment. This structure is, in fact, an emergent phenomenon, irreducible within this level of the system. The real question is how did that emergent structure come about? To consider some possible answers, we need to move from the macroscopic level and look at the microscopic level.

AIR TRANSAT 236

On August 24, 2001, Air Transat 236, an Airbus A330 enroute to Lisbon from Toronto, arrived in truly spectacular fashion on runway 33 at Lajes Airport in the Azore Islands. The airplane had been without fuel or power for the final nineteen minutes of the flight; Captain Robert Piche and First Officer Dirk de Jager had executed a near-perfect deadstick landing in a 370,000-pound airplane on a 10,000-foot runway situated on a rather small island in the middle of an ocean.[45]

How they got there is a remarkable story, but it begins in Air Transat's Montreal maintenance hangar. The right engine on the airplane was making metal, and so an engine replacement was planned, with work beginning at midnight on Friday, August 17 and expected completion by about noon on Sunday. As it happened, the engine being removed, as well as every other Trent engine in the Air Transat fleet, had been modified through two Rolls Royce SBs to accommodate an improved rear hydraulic pump. One SB detailed the new pump, while the second detailed the replacement of three fuel tubes and two hydraulic tubes to accommodate the new pump. The pump replacement was optional; however, if you replaced the pump, you absolutely had to replace the fuel and hydraulic tubes. As it happened, the replacement engine had not been so modified.

The engine change procedure involved removing the accessory package, including the rear hydraulic pump, so that these accessories could be refitted to the replacement engine after it was hung on the wing. The procedure *did not* involve removal of the fuel and hydraulic tubes; those stayed with the engine. When the technicians attempted to fit the newer hydraulic pump to the unmodified replacement engine, they encountered interference with fuel lines. At that point, they discovered the problem.

Several links in the "chain of causality" appeared. To begin with, the replacement engine had not been evaluated for SB compliance when it was received, so there was no information indicating that it was inconsistent with the configuration existent on the rest of the fleet. Next, the technicians doing the work encountered a network problem on the company computer system that prevented them from accessing the SBs. Third, there was no older, pre-modification hydraulic pump available. The only procedurally correct solution was to use the post-modification pump and to replace the fuel and hydraulic tubes; in other words, accomplish the SBs to the letter and thus complete the modification.

But that isn't what they did. In order to fit the newer hydraulic pump, it was only physically necessary to replace the fuel tubes. The older, unmodified hydraulic tubes did not need to be replaced in order to achieve the correct fit of the pump. This work was accomplished without reference to the actual SBs. It was then noted that the older hydraulic line did not have the required clearance between it and the fuel line. This was resolved by physically positioning the hydraulic line such that clearance was achieved and then tightening the "B" nut on the hydraulic line attachment fitting to maintain the clearance.

More holes in the cheese appeared. Nowhere in any of the training manuals or material used by the maintenance technicians was there any discussion of the possibility that hydraulic pressure could easily overcome the forced positioning achieved with the "B" nut tension. Which, ultimately, is exactly what it did, leading to a chafing of the fuel tube that finally opened a rather gaping hole, dumping fuel overboard at a ridiculous rate halfway across the Atlantic.

It is easy to see how this series of events can be drawn historically, with a time vector, so that the links in the chain, or the slices of cheese, are aligned in a linear fashion. If we only look at the chain of causality in this accident, we can very easily default to a version of "capable and competent airman" … in this case, the maintenance technicians … managing to fail in their duty to prevent the airplane from cracking up. Clearly, there were interventions already existent which could have arrested this sequence of events, broken the chain, such as a requirement to read the applicable SBs. So what happened? Why was that intervention unsuccessful?

In the report, the Portuguese authorities elaborated on the history of dialog between the lead technician on the hangar floor, the Rolls Royce tech rep and the Air Transat engine controller, who worked from the airline's Maintenance Control Center. All were involved in extensive discussions about the problem and how to resolve it. What needs to be examined is the quality of the information transfer between the parties and higher level effects of that information transfer.

The lead technician and the engine controller discussed the difficulty with accessing the SBs. The Maintenance Control Center had access to these bulletins via a set of compact discs. The Engine Controller did not use this access, and the lead technician was not aware it existed. The focus of their conversation was the practical aspect of ensuring adequate clearance between fuel and hydraulic lines following installation. The Rolls tech rep phoned in to offer assistance, and he was advised that the fuel lines were being changed. He did not access the SBs and was not aware of the corresponding requirement to change the hydraulic tubes. He offered to visit the hangar, but that offer was not taken up.

As Fromm has said, we are often only able to see part of the system, either the macroscopic *or* the microscopic, the system *or* the environment, but not both. The organized, thought-out standard operating procedure exists on a visible and rational level, easily comprehended through the narrative mode of thought. But the next level of this system exists at a less visible stratum, that of the inter-relationships between the actors, in this case the lead technician, the Engine Controller, and the Rolls tech rep, with some additional agents undoubtedly involved in the form of other members of the technical team. Organizationally, they have defined roles and a defined relationship. The open question is what happens when (a) the hydraulic pump won't fit and (b) the SB cannot be accessed.

Damienne Provitolo et al. have provided a thoughtful examination of how complex human behaviors can arise during and following major crises and disasters, by applying Fromm's model of emergence to various manifestations of human behavior.[46] In particular, she has described a form of Fromm's weak emergence, in which feedback loops exist from the macro level to the micro, as well as from the micro to the macro. In the unstable version of this type of weak emergence, both feedback loops are positive and the emergence of a new structure tends to reinforce the behavior. She uses as an example the spread of panic, in which a few panicking individuals influence other, non-panicking individuals, causing them to panic. Eventually the number of people experiencing panic becomes large enough to reinforce the sense of panic across the entire group. Figure 5.8 shows a somewhat whimsical illustration of this idea, but if you look closely, you will see the fear communicated between the two men in the foreground is beginning to spread to those immediately surrounding them, initiating an emergent panic, while others in the background appear unalarmed as yet. Conversely, a stable form of weak emergence exists when the micro-to-macro feedback is positive, but the macro-micro feedback is negative, regulating the system and preventing unstable accelerations of behavior.

Through the personal perceptions of each other that the Air Transat technical crew held among themselves, the Engine Controller, and the Rolls Tech rep, the ease and clarity of communications and dialog, and the varying levels of technical comprehension, the structure of the relationships was altered from that which would exist with a complete knowledge of the SBs available, which is the structure that the standard operating procedure assumes will exist, to one which still achieved the dominant thought without the SBs. The agreement regarding what technical steps are pragmatically necessary ... changing the fuel tubes ... reinforced this new structure of the relationships between the parties in a positive feedback loop that moved the new structure farther away from the formal structure. Hence, at this less visible level of the human relationships between the technical personnel present, an emergent structure appeared, which is the informal relationship between the technical

FIGURE 5.8 Panic as an emergent structure.

personnel. It is somewhat self-organized, and nothing in the individual communications would predict the change in relationships.

When the structure of collective behavior departs from that of compliance with standard operating procedures, it is actually shifting from a classic rational choice model, in which sensemaking organizes and comprehends all of the necessary information as inputs to a decision, and the decision itself is decoupled from the sensemaking process that generated the information, to a endogeneity model, in which the decision is coupled to the sensemaking process and the output of decision-making serves as an input to sensemaking.[47]

This is a form of what is known as action endogeneity. Endogeneity is generally considered to occur when an output value of a process or calculation alters the value of one or more variables in the calculation; this is itself a form of feedback. For action endogeneity, the process of moving through each step in a course of action alters the available cues, modifying the environment and revising the sensemaking.[48] Figure 5.9 shows an illustration of how this works.

Endogeneity becomes extremely important when evaluating what is known as a self-fulfilling confirmation loop, which is a necessary structural component within any process of action-based problem-solving. Jenny Rudolph and her colleagues have explained the self-fulfilling confirmation loop as a process for receiving, interpreting, and giving weight to cues generated following an action. It can be both constructive and destructive. It is constructive if it slows the abandonment of a particular path of

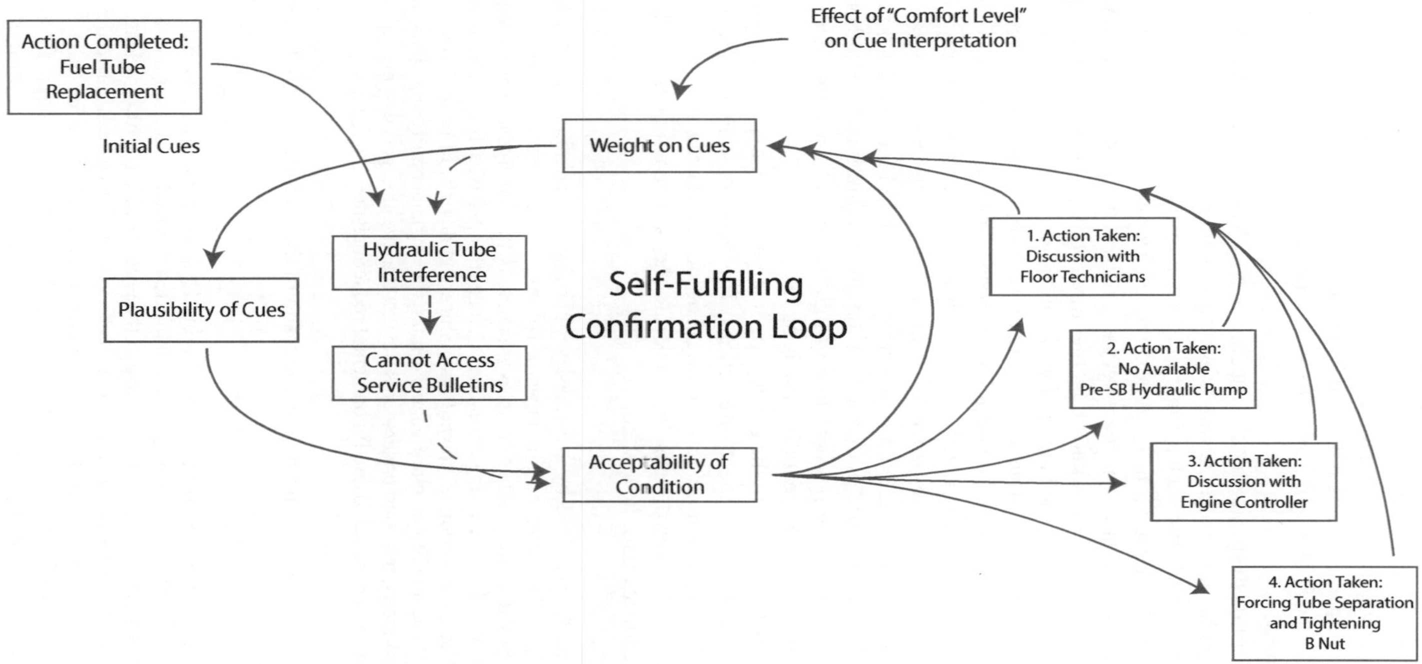

FIGURE 5.9 Action endogeneity on the Air Transat hangar floor at Montreal.

action until sufficient cues have been generated to either confirm or disconfirm that path. It can be destructive if the cues are misinterpreted or incorrectly weighted, thus reinforcing an incorrect path of action. Indeed, the self-fulfilling confirmation loop is particularly vulnerable in situations in which action is constrained or information feedback is delayed.[49] This can lead to fixation on a particular mental model or situational awareness.

So it is important to spend a bit of time considering how cues are generated, interpreted, and weighted, because getting those right is largely within our control and can make the difference between a frustrating day and a devastating day.

In their paper, "Understanding Complexity in the Human Brain", Danielle Bassett and Michael Gazzaniga have written a very extensive survey of the role complexity and emergence play in what is known as the mind-body problem.[50] Simply put, we know where the brain is in the body and we know quite a lot about how it works. We cannot identify the location of what we think of as the mind. We do not have a good deterministic concept of how the physiological properties and functions of the brain lead to consciousness or thought. In their analysis, the brain is a highly complex system comprised of extensive networks, leading, on a higher level of organization, through an emergent process, to consciousness and mental function. They point out an important distinction, the "disconnect in the mind–brain interface: although the material components of the physical brain might be highly decomposable, mental properties seem to be fundamentally indivisible".[51]

These authors then reiterate the concept of emergence occurring at a higher level than the fundamental component interactions. However, as shown in Figure 5.10, they make a very important point by stating that the manifested emergent phenomena can produce a downward causality back into the basic component level, a form of top-down feedback into the system.

> Emergence is characterized by a higher-level phenomenon stemming from a lower system level; that is, emergence is upward. However, an important property of the brain, as opposed to some other complex systems, is that emergent phenomena can feed back to lower levels, causing lower level changes through what is called downward causation. The combination of upward emergence and downward causation suggests a simple bi-directionality or more nuanced mutual complementarity that adds to the complexity of the system, and underscores the fact that the emergence of mental properties cannot be understood using fundamental reductionism.[52]

The Portuguese authorities summed up the result of the dialog between the technical personnel this way:

> When the lead technician contacted the Engine Controller, the Engine Controller's knowledge of the SB and its background comforted the lead technician into feeling that the Engine Controller had a good grasp of the problem at hand. Acknowledgement by the Engine Controller that the fuel tube needed to be replaced confirmed the lead technician's mental model that this was the only requirement for completion of the installation. The confirmation was reinforced by the fact that the Engine Controller was associated with the engineering department, which had the responsibility for resolving unexpected or non-routine maintenance issues.[53]

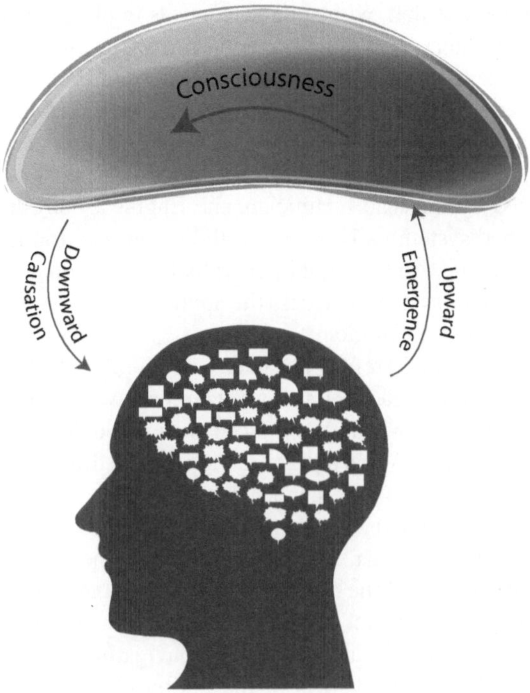

FIGURE 5.10 Emergence through the mind-brain interface.

The emergent phenomenon is the perception of the situation, in this case, is the sense of "comfort" that the lead technician felt when talking this over with the Engine Controller. This sense was not an explicit intent held or imparted by the Engine Controller. Rather, it was a state of mind that is irreducible to the bits of information themselves or to the physiological functions within the lead technician's brain. It is largely influenced by the top-down schedule pressures and roadblocks to standard procedures. This goes directly back to Jerome Bruner's two different landscapes: the landscape of action, in which the participants are the origin of actions, intentions, and goals, and the landscape of consciousness, which describes the participant's thoughts, knowledge, and feelings, or perhaps more importantly, what they do not think, know, or feel.[54]

In the end, there was considerable downward causation, in the form of schedule pressure as well as how that schedule pressure molded the upward emergence that influenced each technician's consciousness, reaching all the way back down to the literal brain functions of the technicians. Due to action endogeneity, moving through the process changed the environment and altered the cues being exchanged, creating an altered relational structure between the personnel involved as well as between each of them individually and the standard operating procedures. Those altered relationships, and the quality of the associated communications, influenced an emergent state of mind that manifested a level of comfort with the actions taken. Ultimately, the emergent phenomena at the macro level was a non-compliant structure of work and inspection that resulted in a serious hazard being accepted.

It is tempting to believe that, with suitable controls in place, the macro-level system organization can intervene on these mental processes and prevent perceptions such as the comfort level the lead technician experienced. Yet those controls were in place. Procedures existed for properly cataloging the status of the replacement engine when it was originally delivered, and certainly procedures existed to read the applicable SBs detailing the hydraulic pump modification. The emergent states of mind experienced by the lead technician, the Engine Controller, and the Rolls tech rep all functioned systemically to bypass those controls. In the case of the lead technician and the Engine Controller, they literally bumped squarely into the failure of the company computer system to access the applicable SBs … and simply created structure to work around this problem.

So how do we correct that? Carefully designed procedural controls can go a long way, but we are still up against the lead technician's "comfort" with the Engine Controller's "good grasp" of the problem when all are faced with schedule pressure and a failed computer data access system. What we are really interested in doing is adapting the self-fulfilling confirmation loop in a manner that provides negative feedback, so that we can regulate the system and try to ensure a stable form of weak emergence.

To think about how to do that, we need to consider the role of what has been referred to as the "drive toward the dominant thought",[55] and its effects on problem-solving. Dan Griffin, an instructor and pilot at American Airlines, introduced the phrase to provide an alternative attitude toward navigating the stress of recurrent training. The main idea is that you have the option of changing your dominant thought and using that to interpret cues and information. A positive attitude allows objective and constructive interpretation, whereas a negative attitude can reinforce incorrect interpretation. One approach fosters learning, the other impedes it. The idea has manifest applications.

The point of the concept behind the phrase "drive toward the dominant thought" is that you have the ability to change the dominant thought, which can go a long way toward ensuring stability as weakly emergent interpretations and behaviors are generated. On the face of it, Griffin's argument seems to be that this conscious decision will lead to complete rationality and the resulting virtue. Implicit in that interpretation is an idea of top-down organization and control over our own thinking that, if we have learned anything from the above discussion, is probably more of a mirage than truth. In fact, what he is really saying with the words "drive toward" is that we must consciously be aware of the self-fulfilling confirmation loop, monitor it for signs of unstable, positive feedback, and apply the drive toward the dominant thought to stabilize that feedback.

The dominant thought in the Air Transat hangar was a mission-oriented response to significant schedule pressure. The altered relationships and comfort with communications generated unstable, positive feedback to the self-fulfilling confirmation loops. What kind of "dominant thought" might have dampened these behaviors, and created some stability in the self-fulfilling confirmation loops? It is important to remember that while each individual's brain is a spectacularly complex system, it is also very much an open system. So, the movement of energy and information must be considered both internally, as between one individual's thought processes, and externally, with respect to how that one person interacts with his or her environment and the other people within it.

THE ECOLOGY OF ACTION

The French philosopher, Edgar Morin, has provided an idea of how to think about this problem at a macro, broad-scale level, which he refers to as the "ecology of action":

> As soon as a person begins any action whatsoever, the action starts to escape from his intentions. It enters into a sphere of interactions and is finally grasped by the environment in a way that may be contrary to the initial intention. So we have to follow the action and try to correct it if it is not too late, or sometimes shoot it down, like NASA exploding a rocket that has veered off course.
>
> Ecology of action means taking into account the complexity it posits, meaning random incidents, chance, initiative, decision, the unexpected, the unforeseen, and awareness of deviations and transformations.[56]

Translated into contemporary thought on complexity, the ecology of action means taking into account the initial conditions, external influences on an open system, local rules, feedback, the progress of consequence, emergent characteristics, and diligently looking for new patterns of self-organization.

Oddly enough, the second half of this accident gives a parallel example of these emergent phenomena and then takes us full circle to the beginning of this chapter.

En route to Lisbon, the problem initially appeared through the unusually low oil temperature and high oil pressure displayed by the right engine, along with a notable decrease in oil quantity. The flight crew contacted the company maintenance control via HF radio and discussed this with maintenance technicians. No one could posit an explanation for these observations.[57]

Shortly thereafter, the crew noted a significant fuel imbalance through an ECAM advisory. To that point, the fuel distribution had been normal. The crew then executed a crossfeed procedure from memory, rather than consulting the QRH. They established fuel flow from the left main tank to the right engine. Consider this in the context of Morin's ecology of action: did the action of crossfeeding escape their original intentions? Was it grasped by the environment, by a sphere of interactions?

As the total fuel load decreased to below the minimum required to continue on the Lisbon, the crew asked the lead flight attendant to inspect the aircraft exterior from the cabin windows, in order to see whether a fuel leak could be observed. No leak was seen. The crew then reconfigured the crossfeed so that the right main tank was feeding both engines, in an effort to utilize the right main tank fuel before it was completely exhausted. At this point, there was a clear opportunity to enter the QRH and execute one of two fuel leak procedures, either one of which would have stopped the leak.

Subsequently, in response to a suggestion from the maintenance controller that a leak may exist in the left fuel tank or engine, the crew briefly reconfigured the crossfeed once again to feed both engines from the left main tank.[58]

Indeed, this crew spent a lot of time in conversation with maintenance control via a radio link. While the Portuguese authorities considered much of this conversation to be non-pertinent, in fact the crew was adhering to the existing model of CRM during what they initially perceived as an abnormal, but not immediately critical, situation.

Figure 5.11 shows the role of action endogeneity in this case, as well as the effect of the ecology of action. In order to equip the self-fulfilling confirmation loop with

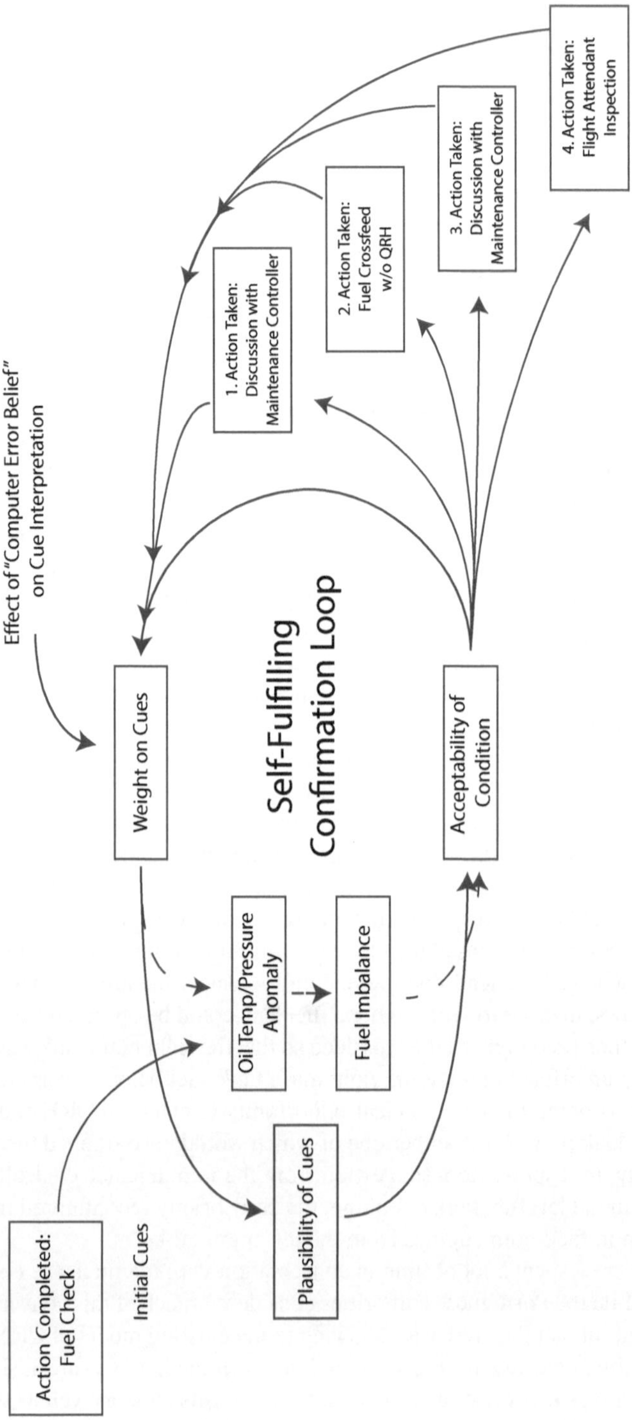

FIGURE 5.11 Action endogeneity in the Air Transat cockpit.

negative feedback, the crew has to be aware of the unexpected, the unforeseen, and most certainly of deviations and transformations.

Yet what is interesting is the departure from procedure, not altogether unlike their counterparts in the hangar at Montreal. The inputs to this situation were widely dispersed; oil parameter discrepancies (but all within acceptable ranges), fuel imbalance, loss of total fuel, no evidence of a visible fuel leak, no other engine problems of discrepancies, and in particular, the suddenness of the onset and development of the situation. All of these inputs engaged a self-fulfilling confirmation loop that, partly through action endogeneity, strongly influenced the perceptions experienced. The crew really did not believe they had an actual fuel leak, even hypothesizing that a computer malfunction may account for the wide range of seemingly unrelated indications.[59]

And this does indeed bring us full circle back to the beginning. The upset experienced by the Qantas 72 crew was a genuinely emergent characteristic appearing at the macro-level of a complex physical system that had faulted at some unidentifiable micro level. On the other hand, the massive fuel leak experienced by the Air Transat crew was linear in rather spectacular fashion. Yet the broad perception of the A330 as a complex system, fully capable of non-linear cause-and-effect at the cockpit level, appears to have created a top-down feedback to the crew's perceptions of the situation; hence the suspicion of a computer malfunction explaining the observations, and quite probably, the multiple conversations with maintenance controllers. At the mind-brain interface described by Bassett and Gazzaniga, this feedback, in conjunction with the multiple inputs from aircraft systems and maintenance control, the absence of an observable fuel leak, and probably several other factors, led to an emergent interpretation of the situation that denied the obvious solution of executing one of two possible fuel leak procedures contained in the QRH, either one of which would have arrested the problem and allowed a single-engine landing at Lajes.

Ask yourself this: what would the crew have done had the same leak occurred on a classic Boeing 747 equipped with steam gauges and a flight engineer?

NOTES

1. Australian Transport Safety Bureau, AO-2008-070, In-Flight Upset 154 km West of Learmonth, WA, October 7, 2008, VH-QPA Airbus A330-303.
2. Ibid., 213.
3. Ibid., 200.
4. "We ought then to regard the present state of the universe as the effect of its anterior state and as the cause of the one which is to follow. Given for one instant an intelligence which could comprehend all the forces by which nature is animated and the respective situation of the beings who compose it – an intelligence sufficiently vast to submit these data to analysis – it would embrace in the same formula the movements of the greatest bodies of the universe and those of the lightest atom; for it, nothing would be uncertain and the future, as the past, would be present to its eyes. The human mind offers, in the perfection which it has been able to give to astronomy, a feeble idea of this intelligence". Pierre Simon Laplace. *A Philosophical Essay on Probabilities*. John Wiley and Sons. Chronicon Kindle Edition, 1902.
5. Thomas Aquinas, *Summa Theologica*, translated by Fathers of the English Dominican Province, e-artnow Kindle Edition, 2013, IIa-IIae, q.49, a.5, ad.obj.2.

6. Sidney Dekker, "What Is Rational About Killing a Patient with an Overdose? Enlightenment, Continental Philosophy and the Role of the Human Subject in System Failure", *Ergonomics*, Vol. 54, No. 8, August 2011, 681.
7. Australian Transport Safety Bureau, 202.
8. National Transportation Safety Board, AAR-90/06, United Airlines Flight 232, McDonnell Douglas DC-10-40, Sioux Gateway Airport, Sioux City, Iowa, July 19, 1989, 78.
9. National Transportation Safety Board, AAR-98/01, Uncontained Engine Failure, Delta Air Lines Flight 1288, McDonnell Douglas MD-88, N927DA, Pensacola, Florida, July 6, 1996, 74.
10. Australian Transport Safety Bureau, 199.
11. Ibid., 135.
12. Federal Aviation Administration, Advisory Circular 00-6A, "Aviation Weather for Pilots and Flight Operations Personnel", 1975, 92.
13. Federal Aviation Administration, *Aeronautical Information Manual, Official Guide to Basic Flight Information and ATC Procedures*, December, 2015, PCGI-1.
14. Jerome Bruner, "The Reality of Fiction", *McGill Journal of Education*, Vol. 40, No. 1, Winter 2005, 58.
15. Australian Transport Safety Bureau, 202.
16. Henri Poincare, *Science and Method*, Thomas Nelson and Sons, 1914, Digitized by the Internet Archive, 2015, https://archive.org/details/b21974123, 68.
17. Nassim Nicholas Taleb, *The Black Swan: The Impact of the Highly Improbable*, Random House Publishing Group. Kindle Edition, 2010.
18. Poincare, 68.
19. Ibid.
20. Lewis Wolpert, Alison Richards, *A Passion for Science,* Oxford University Press, 1988, 42.
21. John H. Holland, *Complexity: A Very Short Introduction* (Very Short Introductions), Oxford University Press. Kindle Edition, 2014, 5.
22. Melanie Mitchell, *Complexity, A Guided Tour*, Oxford University Press. Kindle Edition, 2009, 13.
23. Jennifer Oullette, "A Fundamental Theory to Model the Mind", *Quanta Magazine*, April 3, 2014.
24. Holland, 6.
25. Jochem Fromm, "Types and Forms of Emergence", Cornell University arXiv platform, June, 2005 https://arxiv.org/abs/nlin/0506028v1.
26. Ibid.
27. Australian Transport Safety Bureau, 202.
28. Fromm.
29. Ibid.
30. Michael Collins, *Carrying the Fire, An Astronaut's Journey*, Farrar, Strauss and Giroux, 1974, 231–238.
31. Sidney Dekker, *Drift into Failure: From Hunting Broken Components to Understanding Complex Systems*, Ashgate Publishing Ltd., Kindle Edition, 2011, 140.
32. Australian Transport Safety Bureau, 259.
33. Poincare, 68.
34. Fromm.
35. National Transportation Safety Board, Aviation Incident Final Report, NTSB No. SEA00IA019, N772AS, Boeing 737-400, Portland Oregon, November 11, 1999.
36. National Transportation Safety Board, Aviation Incident Final Report, NTSB No. LAX01IA066, N735D, Lockheed L-1011-385-1, Honolulu Hawaii, December 29, 2000.
37. National Transportation Safety Board, Safety Recommendations A-06-29 through -35, March 30, 2006.
38. Fromm.

39. Edward N. Lorenz, "Deterministic Non-Periodic Flow", *Journal of Atmospheric Sciences*, Vol. 20, March, 1963.
40. Fromm.
41. Federal Aviation Administration, *Aeronautical Information Manual.*
42. Fromm.
43. National Weather Service Chicago, Forecast Discussion, n.d., accessed June, 2019.
44. Rebecca M. Cintineo, David J. Stensrud, "On the Predictability of Supercell Thunderstorm Evolution", American Meteorological Society, July 2013, doi: 10.1175/Jas-D-12-0166.1.
45. Government of Portugal, Ministério Das Obras Públicas, Transportes E Comunicações, Gabinete De Prevenção E Investigação De Acidentes Com Aeronaves, Accident Investigation Final Report, All Engines-Out Landing due to Fuel Exhaustion, Air Transat Airbus A330-243, C-GITS, Lajes, Azores, Portugal, August 24, 2001.
46. Damienne Provitolo, Edwige Dubos-Paillard, Jean-Pierre Muller, "Emergent Human Behaviour During a Disaster: Thematic Versus Complex Systems Approaches", Proceedings of EPNACS 2011 within ECCS'11 Emergent Properties in Natural and Artificial Complex Systems Vienna, Austria, September 15, 2011.
47. Jenny W. Rudolph, J. Bradley Morrison, John S. Carroll, "The Dynamics of Action-Oriented Problem Solving: Linking Interpretation and Choice", *Academy of Management Review*, Vol. 34, No. 4, 2009, 749.
48. Ibid., 748.
49. Ibid., 750.
50. Danielle S. Bassett, Michael S. Gazzaniga, "Understanding Complexity in the Human Brain", Trends in Cognitive Sciences, Vol. 15, No. 5, May 2011, 200–209, doi: 10.1016/j.tics.2011.03.006.
51. "Perhaps most simply, emergence – of consciousness or otherwise – in the human brain can be thought of as characterizing the interaction between two broad levels: the mind and the physical brain. To visualize this dichotomy, imagine that you are walking … through a mill. Consider that you can blow the mill up in size such that all components are magnified and you can walk among them. All that you find are mechanical components that push against each other but there is little if any trace of the function of the whole mill represented at this level. This analogy points to an important disconnect in the mind–brain interface: although the material components of the physical brain might be highly decomposable, mental properties seem to be fundamentally indivisible". Danielle S. Bassett, Michael S. Gazzaniga, "Understanding Complexity in the Human Brain", Trends in Cognitive Sciences, Vol. 15, No. 5, May 2011, 200–209, doi: 10.1016/j.tics.2011.03.006.
52. Ibid.
53. Government of Portugal, 71.
54. Jerome Bruner, *Actual Minds, Possible Worlds*, Harvard University Press, 1986.
55. Personal Communication, Captain Dan Griffin, American Airlines.
56. Edgar Morin, *Seven Complex Lessons in Education for the Future*, United Nations Cultural Organization, November 1999, 45.
57. Government of Portugal, 74.
58. Ibid., 57.
59. Ibid., 58.

6 Mindfulness

On the night of January 27, 1986, a telephone conference was held between engineers and managers at Morton Thiokol, the manufacturer of the space shuttle solid rocket boosters, and their counterparts at NASA Marshall Space Flight Center and the Kennedy Launch Complex. The purpose was to discuss the safety of proceeding with the scheduled launch of the space shuttle Challenger the following morning. Several Thiokol engineers, having at least a gut sense that the cold temperatures forecast for the next morning were placing them deep into uncharted territory, deep into a realm of average risk, had exercised a rich awareness of discriminatory detail … mindfulness … and had raised objections over the plan to launch.[1]

Mindfulness reverberates throughout the herd. The zebra who is farthest from the zebra who actually smells the lion doesn't have to trot over and check the odor himself; his own adherence to mindfulness will detect his fellow zebra's reactions when sensing the odor, and that will be enough for him. But our zebra analogy has to stop here. The zebra doesn't have any emotional or psychological issues that interfere with his mindfulness. He has no particular opinion of himself, no pride, and no self-esteem. He doesn't see himself in any manner whatsoever; he wouldn't know what a mirror was, and he quite probably would not recognize himself if he saw his reflection in one. He feels no guilt, no embarrassment, and no inadequacy in any way. He has only one abiding concern: staying alive. If he suddenly realizes that he has, up-until-exactly-right now, failed to detect a lion crouching in the grass, it is no skin off his nose to accept this new information immediately and act accordingly. Zero to sixty, hopefully in less time than it takes the lion.

We are not so simple. There are a couple of obvious impediments to mindfulness, such as fatigue, distraction, and sensory function, that is, illusions. However, two quite pervasive impediments to mindfulness are more subtle. The first can arise when the discriminatory detail of which we have just become richly aware happens to conflict with our own previously formed opinions, beliefs, or plans. This conflict is fertile ground for Leon Festinger's idea of cognitive dissonance, the inner psychological tension that is generated when faced with two equally held but inconsistent beliefs, opinions, or attitudes. The second occurs when the value of the discriminatory detail has been degraded and ultimately ignored through the misinterpretation of experience. This is known as the normalization of deviance.

COGNITIVE DISSONANCE

On August 20, 2011, FirstAir 6560, a Boeing 737-200, crashed at Resolute Bay when the crew failed to capture the localizer and configure the aircraft during the final approach. For reasons never determined with certainty, the aircraft compass system had drifted about 17° during the descent. The autoflight system initially captured the localizer signal, but the aircraft passed through the beam and it was necessary

DOI: 10.1201/9781003369677-6

to re-intercept the localizer from the other side. At this point, the autopilot dropped out of the localizer capture mode into a heading hold mode, also for reasons never determined with certainty. However, it entered heading hold on an indicated heading that appeared adequate to re-capture the localizer. This was misleading, since the compass systems were in error, and the true heading was actually slightly divergent from the final approach course. Neither the captain, who was flying, nor the first officer had any inkling of the compass problem, and they did not appear to recognize the autopilot mode problem; but the first officer did understand that the aircraft was well to the right of the localizer with a full deflection of the localizer needle on the navigation display. He voiced this to the captain in several ways; curiously, none of his attempts were effective.[2]

Leon Festinger's original concept of cognitive dissonance can be envisioned when you consider, for example, the guy dressed up like Tutankhamun who is sitting cross-legged about eighteen inches above the sidewalk in Barcelona, with absolutely no visible means of support. This will drive most people to distraction, generating a palpable psychological tension, because it is just so wrong, and ultimately, we are pushed to find a resolution of some kind. With a little study, of course, you soon figure out that he does have a means of support and that the whole thing is an illusion. Once you understand this, the tension evaporates; there is no more dissonance. It is easy to resolve, although the first couple of minutes staring at him can be a little challenging. For this service, of course, he would appreciate a small financial contribution.

However, a more stressful cognitive dissonance is encountered when the conflict cannot be resolved through accessible and concrete explanation, but rather requires that we adjust a belief, opinion, or attitude. We encounter these kinds of dissonance every day. When I was a kid in Manhattan, it was common to see a police car calmly cruise right through a red light, without any indication of urgency. Most of us try to watch what we eat, but most of us are also attracted to foods we probably don't need to consume in any quantity. Many of us find homeless people unattractive and quite possibly repulsive, while we also believe that we should care for them. Of late, we have found it increasingly difficult to believe that a political leader for whom we did not vote might actually be competent. This creates a tension as shown in Figure 6.1.

The tension can become more pronounced with the strength of our beliefs in the conflicting ideas. How was it possible for Thomas Jefferson to have an affair with Sally Hemmings and still be Thomas Jefferson? For that matter, how was it possible for Jefferson to be Jefferson while *owning* Sally Hemmings? How was it possible for Lincoln, the Great Emancipator, to believe that blacks were naturally inferior to

FIGURE 6.1 Cognitive dissonance has its roots in conflict between facts and beliefs.

whites, and so argue in his debates with Stephen Douglas? How can we love our dogs and cats while subjecting deer to agonizing death by hunting them with bows and arrows? How can a capable and competent aviator ever allow his ship to crack up?

These dissonances represent what psychologists refer to as the belief disconfirmation paradigm. Festinger described the paradigm this way:

> Dissonance and consonance are relations among cognitions – that is, among opinions, beliefs, knowledge of the environment, and knowledge of one's own actions and feelings. Two opinions, or beliefs, or items of knowledge are dissonant with each other if they do not fit together – that is, if they are inconsistent, or if, considering only the particular two items, one does not follow from the other. For example, a cigarette smoker who believes that smoking is bad for his health has an opinion that is dissonant with the knowledge that he is continuing to smoke. He may have many other opinions, beliefs, or items of knowledge that are consonant with continuing to smoke but the dissonance nevertheless exists too. Dissonance produces discomfort and, correspondingly, there will arise pressures to reduce or eliminate the dissonance. Attempts to reduce dissonance represent the observable manifestations that dissonance exists. Such attempts may take any or all of three forms. The person may try to change one or more of the beliefs, opinions, or behaviors involved in the dissonance; to acquire new information or beliefs that will increase the existing consonance and thus cause the total dissonance to be reduced; or to forget or reduce the importance of those cognitions that are in a dissonant relationship.[3]

A second, nearly omnipresent aspect is the role played by the sense of self-esteem. We normally maintain a favorable impression of ourselves and tend to disdain error because it detracts from that impression. We do not typically perceive ourselves as incompetent, or prone to mistakes or errors, and we certainly do not get up in the morning, look at ourselves in the mirror, and see a person we believe is likely to cause a hundred or more fatalities due to some oversight. A sizeable body of cognitive dissonance theory focuses on self-consistency and how we respond to specific aversive consequences that result from the use of free will. For example, if we make a decision that we then suddenly discover has placed the aircraft in an undesirable state, we will experience a powerful dissonance. How could I have been that wrong? How could I have allowed this or that error, or series of errors, to have taken place? This is not consistent with who I am! This type of dissonance is made all the more problematic if we perceive that the situation was foreseeable.

How we respond to the inner drive to reduce dissonance may not always be as rational and pragmatic as we may like. We need to be careful of this; it is easy to intellectualize the problem, as nearly all of us would admit that it is certainly possible for any one of us to screw the pooch, so-to-speak … there, but for the grace of God, go I … but intellectualizing it and fully integrating it into our worldview are two different things. We are very likely to reject the latter simply because it requires us to live with round-the-clock … persistent … vulnerability.

A third aspect of cognitive dissonance is known as the "action-based model of dissonance". Eddie Harmon-Jones, who has developed this idea, defines it thus:

> "The model … proposes that dissonance between cognitions evokes an aversive state because it has the potential to interfere with effective and un-conflicted action".[4]

It is fairly well-understood that an annoying degree of cognitive dissonance is generated when a particular decision features options that are all closely ranked in favorability, making a choice difficult. In cases featuring one obviously best option, dissonance is at a minimum; the choice is easy. But in cases involving multiple and generally equivalent options, dissonance may be more powerful. Consider this hypothetical quandary: to whom would you like to give the one extra ticket for a seven-day Caribbean cruise that you mysteriously inherited? Let's say your choices are limited to either all of the contestants in this year's Miss America contest, or the starting lineup of the New York Yankees … depending on your particular interest. It is also well-documented in cognitive dissonance research that, once a difficult decision has been made, the individual will tend to reduce the dissonance by spreading the alternatives, widening the pre-decision ranking so as to favor the choice that has been made … particularly once the cruise ship has left port. This is part of what is known as the free-choice paradigm.

Harmon-Jones' point is that once this difficult decision has been made, the cognitive dissonance arising from the proximity of roughly equivalent options can interfere with following through on the commitment to that decision. In this "action-oriented" state, immediately following the decision, the reduction of that cognitive dissonance will be at a premium in order to take "effective and un-conflicted" action and implement the decision.

Needless to say, a pilot spends the bulk of his or her day managing effective and un-conflicted action. The potential for an encounter with cognitive dissonance is inherent in Weick's description of mindfulness, that is, things that "disconfirm, are unpleasant, feel uncertain, seem possible, are implicit, and are contested".[5] The very act of practicing mindfulness is likely to reveal dissonant cognitions and generate psychological tension. Whether it be the disconfirmation of beliefs, challenges to self-affirmation, or interference with effective and un-conflicted action, the prospect of encountering cognitive dissonance and the inner stress that it creates can easily discourage mindfulness.

The crew of Air Transat 236, who experienced fuel exhaustion over the Atlantic, readily admitted that they did not really believe that they had fuel leak until the right engine flamed out. Up to that point, while they had considered the possibility of a fuel leak, they thought it much more probable that they were experiencing a computer error or indication malfunction of some sort.[6] There were, in their estimation, no other indications of an actual fuel leak and even a visual inspection by the cabin crew revealed nothing. The Portuguese authorities, in their comprehensive report on the subsequent deadstick landing at Lajes, stated that

> Confirmation bias can have such a strong impact that once individuals have developed a mental model of a problem space, and they have confirmed their model, it becomes very difficult to let go of the model, even in the light of contradictory information.[7]

They also pointed out that

> Additionally, acting upon one's beliefs can also increase the psychological costs or "dissonance" involved in changing one's beliefs.[8]

In the FirstAir case at Resolute Bay, the cognitive dissonance was almost as pure as Tutankhamun in Barcelona. The compass heading and flight directors showed a perfectly acceptable resolution to the problem; the localizer needle and GPS position showed that, in point of fact, the problem is not resolving at all. At this point on any flight, the flight crew has very much entered an "action-oriented" state of mind; this is particularly true of the pilot flying the airplane. Absent any awareness of a technical explanation for the divergent indications, the pilot flying will drive toward effective and un-conflicted action, jettisoning cognitive dissonance along the way as necessary.

And so the captain did. Early in the maneuver, he quite likely believed the situation was resolving, since the indicated heading and wings level flight director indications were consistent with an autopilot capture maneuver. But the aversive state would not dissipate, because the aircraft position was not improving. Less than forty-five seconds before impacting the ground, he stated that a similar situation had happened to another crew previously, as if that somehow made the unstable approach with full needle deflections acceptable.[9]

The Transportation Safety Board of Canada added another dimension to this when they pointed out the effects of increased stress and the resulting narrowing of attention. They said that

> Attentional narrowing can lead to confirmation bias, which causes people to seek out cues that support the desired course of action, to the possible exclusion of critical cues that may support an alternate, less desirable hypothesis.[10]

I think this is close but not entirely correct. A better way of saying this is to say that people seek out cues that can be used to reduce cognitive dissonance, in order to clear a path for effective and un-conflicted action. In an action-oriented state, there is little interest in resolving conflicted action; with attentional narrowing, there is almost no residual attention left over for the mindfulness that would identify and highlight the conflict. The pilot monitoring is less action-oriented and consequently in a much better position to deploy mindfulness, which leads to the question of how best to orient the authority gradient in the cockpit in order to closely couple mindfulness with decision-making.

One of the most interesting aspects of cognitive dissonance theory is the tendency to run a mathematical ratio in one's mind, comparing the number of dissonant thoughts with the number of consonant thoughts. This ratio looks like this:

$$\frac{D}{D+C}$$

where D is the sum of cognitions that are dissonant, or disagree, with a particular cognition and C is the sum of cognitions that agree with the particular cognition. Each cognition is weighted for importance.[11]

Consider the effect of this ratio in the Air Transat case. What was this ratio before the lead flight attendant went to look for visual evidence of a fuel leaks? When she reported seeing no evidence, what did that do to the ratio? It added a consonant

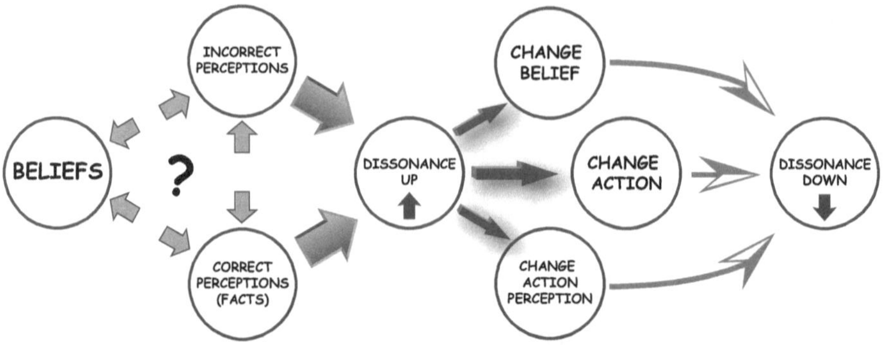

FIGURE 6.2 A diagram of the cognitive dissonance resolution model.

thought to the denominator and reduced the cognitive dissonance. Clearly, sending the flight attendant back to look for a leak was a good use of resource management and could provide valuable information. But the dissonance ratio is really a ratio of Weick's description of mindfulness, that is, a ratio between things that disconfirm and feel uncomfortable over things that disconfirm and feel uncomfortable *plus things that confirm and feel comfortable*. Things that confirm a belief reduce cognitive dissonance. However, the absence of evidence is not evidence of absence. We have to consider all of what we know and believe in any given situation carefully and understand how those beliefs influence our ratio of cognitive dissonance. Figure 6.2 illustrates this process.

Although the old fable of the fox and the grapes represents a traditional conception of cognitive dissonance, one has to be cautious not to underestimate the sophistication and subtlety of the mechanisms that enable dissonance reduction. As I said earlier, the zebra has none of this problem, so raw mindfulness … bolting at the first sight of a lion's ear twitching in the grass … works pretty well for him. When we humans first spot the lion's ear twitching in the grass, we need to be prepared for the rapid onset of psychological tension that may accompany that little tidbit of information … starting with the inevitable first salvo from that part of our brain that wants to know if it really is a lion's ear. How could that actually be a lion? How could you let that lion get so close? Is that really a lion's ear? You've screwed up! Don't overreact, now! As we become aware of that tension, we need to consciously, directly, and decisively ignore the dissonance and focus immediately on the discriminatory detail, become richly aware, and act.

NORMALIZATION OF DEVIANCE

The concept known as normalization of deviance was developed by the sociologist Diane Vaughan during her research into the Challenger disaster of 1986. The actual words used … normalization of deviance … are linguistically adaptable to a range of ideas, in part because they bring along a suggestion of movement from the particular toward the amalgamated, and implicit in that movement is a relaxation of discipline. Indeed, there are at least a couple of applications of this term that really follow different definitions, probably because it is just such an attractive term to use in describing a variety of behaviors. That said, the normalization of deviance, as originally defined by Vaughan, is not a self-interested behavior. It is important to be careful in this definition or we'll miss the point.

Normalization of deviance begins with the process of constructing risk through the interpretation of experience. The interpretation of experience is an explanatory process; we look for explanations of how the experience came to be. In logic, this employs a set of thought processes known as abductive reasoning. These processes are used almost universally to construct hypotheses, ranging from why there is a puddle of oil on the garage floor under my truck all the way out to gravitational waves. Along with inductive reasoning, abductive reasoning forms a body of thought described collectively as ampliative. The term ampliative, within the field of logic, means that both inductive and abductive thinking will generate conclusions that can be extended beyond the limits of the original information. This is the opposite of deductive reasoning, which generates a conclusion valid *only* for the original information.

The man who spent the better part of the latter 19th and early 20th century shaping the ideas behind abductive reasoning, Charles Sanders Peirce, put it this way:

> The surprising fact, C, is observed.
> But if A were true, C would be a matter of course.
> Hence, there is reason to suspect that A is true.[12]

Peirce was not suggesting that "A" is the only possible explanation for "C". Rather, he is ambiguous, arguing at various times that the human mind has a unique power of guessing right, or that humans possess a certain insight that is not quite strong enough to ensure that we are more often right than wrong, but is strong enough to ensure that we are not overwhelmingly more often wrong than right.[13] Ultimately, he argues that the associated inferences arise by controlling our thoughts, actually aligning them with a reasonableness derived from experience.[14] Essentially, this is an educated guess.

The intersection between Peirce's formulation of this idea and the narrative mode of thought described by Bruner is captured by the "surprising fact", which Bruner refers to as the peripeteia, or a departure from canonical order. Thus, abductive reasoning can be used to backfill the narrative, to build a story and make the surprising fact comprehensible.

In no case is an ampliative conclusion either universally or necessarily true. Newton's laws of motion extend to all bodies and actions … until you get to quantum mechanics. An inductive projection that all swans are white, based on years of experience seeing only white swans, is valid until you travel to other parts of the world and encounter a black swan. A turkey's conclusion that, since it has been fed, watered, and sheltered for one hundred days, nothing different will occur on the one hundred and first day is valid until Thanksgiving.

Yet ampliative reasoning is the backbone of science and engineering. All scientific theory is derived through forms of ampliative reasoning, both inductive and abductive. Conversely, all mathematical theorems are derived through deduction.

To the extent that engineering relies on science, it is dependent on ampliative thought. To the extent that it relies on mathematical models, engineering can easily appear to be predicated on deductive thought.

When a mathematician develops a general proof of a theorem, the theorem is deductively bulletproof as long as the original premises remain true. For example, the Pythagorean Theorem is proven in the general case as long as we are talking about two-dimensional Euclidean geometry and as long as one angle is ninety degrees. If you try to use it in three-dimensional cases, you get in trouble. If the angle is a fraction off ninety, the dimension of the triangle's hypotenuse will not be correct. But for all triangles for which the premises are true, the conclusion will always be true.

Ampliative reasoning is different. Way back when the Scottish philosopher, David Hume, first discussed the method of induction, he described it as a method for claiming

> "that instances, of which we have had no experience, must resemble those, of which we have had experience, and that the course of nature continues always uniformly the same".[15]

While the ampliative characteristic allows an inductive projection … the next swan I see is going to be white … the limitations of ampliative reasoning must be borne in mind. Hume pointed out that

> … it implies no contradiction that the course of nature may change, and that an object seemingly like those which we have experienced, may be attended with different or contrary effects.[16]

Gerhard Minnameier explains this in a more contemporary manner, saying that

> induction can only establish truth relative to the current state of affairs. It can never be excluded that future evidence may challenge a currently well-established theory.[17]

Peirce described the way all three forms of reasoning work in a curious triangular relationship, described by Minnameier as the inferential triad, in order to confirm an abductively generated hypothesis. Figure 6.3 is a diagram of the inferential triad. If we begin with one of Peirce's surprising facts, we then back up to construct an explanatory hypothesis for how that surprising fact came to be … backfilling a narrative as we go. Once we have the hypothesis defined, we can then evaluate the hypothesis in the general case and extract some deductively reasoned conclusions from that general case. In other words, in all cases in which the hypothesis is true, certain additional conclusions are generated necessarily through deductive reasoning; these conditions must be true if the hypothesis is true. From this position, we can construct a test, or experiment, to see whether the deductively generated new conditions are actually true in the case that produced our surprising fact. If so, voila! We have a working, albeit unproven, hypothesis. If not, discard that hypothesis, return to the surprising fact with more coffee, and start over.

Peirce's memorable example of how this might work in everyday, non-scientific inquiry is the case of the Catholic priest. Suppose we see a man dressed as a man

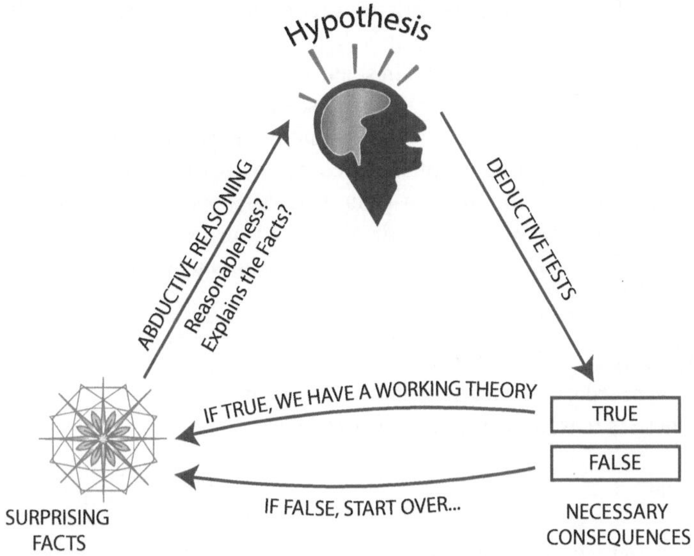

FIGURE 6.3 The inferential triad.

of the cloth, but we have no idea what his religion might be. We suspect that he is a Catholic priest. This is an abductive hypothesis so far … a guess. Now, deductively, we know that all Catholic priests should speak and understand Latin (bear in mind that he posed this example at the turn of the 20th century). If we speak to the man in Latin, his comprehension of what we have said should confirm our hypothesis. So, we construct a test; we casually speak to the man in Latin. It is then clear that he comprehends our statement. But what weight do we place on this information? Does it prove that he is a Catholic priest? No; clearly it supports the hypothesis, but we would need more experiments to reach a more robust conclusion. There are obviously other people in the world, aside from Catholic priests, who speak and understand Latin.[18]

When I set out to locate the source of the oil on the garage floor under my truck, I will probably have a couple of abductively reasoned hypotheses already in my head before I even put on my coveralls. The oil drain plug isn't tight. The rear crankshaft seal is leaking. These hypotheses are explanatory but they are also ampliative. I've never worked on this engine before, but I have worked on many others, and I know from that experience that the oil drain plug may be the problem. I don't need to know that this particular oil drain plug may leak. I can ampliatively reason from my previous experience that it will. I will obviously look for the path of the leaking oil and check the drain plug for snugness.

But to Minnameier's point, one of my favorite aspects of auto maintenance is that once you have found evidence supporting your hypothesis … the oil drain plug turns out to be slightly loose … it may take days before you realize that the discovered evidence has absolutely nothing to do with the actual problem. The oil drain plug was loose, I tightened it, and the next day I didn't see any fresh oil. But a week later

I did, because, of course, the leak actually is coming from the rear seal. Such is the pitfall of an inductively verified abductive hypothesis.

When a theory is considered valid in science, an entirely different situation exists from the work done by the mathematician. A scientific theory is never proven in the general case. It is only proven to the extent that it has yet to be falsified. The word falsified, and the theory of falsification developed by Karl Popper, carry an implicit meaning: instead of proving an idea to be true, let's see if we can prove it to be untrue.[19] If we cannot prove it untrue, then, for the time being, it remains true.

Falsification can be illustrated using Peirce's original formulation of abductive reasoning:

> The surprising fact, C, is observed.
> But if A were true, C would be a matter of course.
> Hence, there is reason to suspect that A is true.[20]

If we now add,

> The surprising fact, D, is observed.
> But if A were true, D could not be true.
> Hence, A has been falsified as an explanation of C.

The subsequent surprising fact, D, represents some of Minnameier's future evidence. In other words, my oil leak is coming from the drain plug, until I snug up the drain plug and subsequently find more oil on the floor. Depending on the evidence, we may completely falsify the hypothesis or we may limit its applicability.

These thought processes are how we come to explain and understand our experience. This is how we troubleshoot, how we problem solve. In its simplest form, abductive reasoning has been described as taking a guess, but it is really an educated guess, and in many cases, it is probably an intuitive guess. Yet with Popper's idea of falsification, we come to a pivotal fork in the road, at which the explanation of experience and the construction of risk diverge completely.

Following one fork, the explanatory interpretation of experience can be presumed true until it is falsified. If and when it is falsified, we simply return to the drawing board and start over on a new, better line of abductive explanation. Following the other fork, the construction of risk *cannot* be presumed true until it is falsified. The falsification of a hypothesis of risk is catastrophic. Further, while we may interpret experience solely on the basis of the current state of affairs, we cannot construct risk on the same basis. This would be akin to only considering credible unknown events to which we have assigned a probability of one … in other words, specific risk … known-knowns. Rather, in order to protect against the consequences of falsification, we need to construct risk on the basis of both specific risk and the premise that "it can never be excluded that future evidence may challenge a currently well-established theory". The future evidence that Minnameier refers to is borne out of the "averaged estimated probabilities of all unknown events" … probabilistic risk … as well as the occasional black swan.

Engineers at NASA and at Morton Thiokol, the manufacturer of the solid rocket boosters, had early concerns about the erosion of the O-rings that seal the joints between the booster's segments. Evidence of this erosion was necessarily discovered after the fact; thus, in each case of erosion, a successful launch had been achieved. The obvious solution to this problem involved completely re-designing and re-manufacturing the solid booster joint. Yet, in each instance, extensive engineering study led to specific conclusions about distinct and specific causes of each case of erosion … conclusions that were really more hypotheses than proven causalities, but each of which conformed to the definition of a specific risk … followed by the implementation of steps to correct each specific problem. A pattern developed; launches subsequent to a particular corrective action would go smoothly and not exhibit O-ring erosion. Then it would crop up again, under conditions somewhat different from the previous erosion cases, and the cycle would repeat. Each of these events contributed to the compilation of an experience base, which bounded all of the historical data from previous launches. As experience with the erosion problem was gained, the failure of any instance of erosion to compromise the O-ring led to the belief that the calculated safety margin had functioned properly in each event. Further study of the booster joint behavior finally led to the conclusion that the erosion phenomenon was self-limiting, that is, that the conditions permitting it to occur were transient and could not last long enough for an O-ring to be breached. Eventually, a primary seal did fail; but the secondary O-ring did not fail, reinforcing the perception of reliable redundancy. Although the original specification had not allowed for any O-ring erosion, primary or secondary, by the time of the Challenger disaster, the erosion was expected and tolerated, indeed even induced through the high-pressure seal tests that were done prior to launch.

Vaughan's extensive work on this event focuses very appropriately on the systemic nature of the O-ring story, particularly the role that bureaucracy, politics, communication, and technical-excellence-no-longer-funded had in the accident. But at its core, there is an individual aspect to the normalization of deviance. Individual engineers working directly with the O-ring problem gained experience with each post-launch discovery of a compromise. Each engineering analysis of each O-ring failure represented isolated pieces of knowledge, understood in the narrative mode of thought, that resolved isolated specific risks and outlined a current state of affairs. When the primary O-ring failed to seal, something was learned about the process of pressurizing the seals after joint assembly, through the story that described that process. When the putty consistency was varied, something was learned about the role that the putty played in sealing the joint … by means of the story that described the putty's behavior. When the pre-launch pressurization of the O-ring chamber was conducted using a lower pressure to minimize the creation of gas paths within the putty, something was learned about the value of a higher test pressure, also through narrative reconstruction.

Each failure identified in the shuttle solid booster joints was processed through Peirce's inferential triad. An abductive hypothesis was developed based on the evidence (if A were true, C would be a matter of course). With this established, some tests or experiments were devised, based on the omission or inclusion of the various observations that suggested the hypothesis in the first place, to demonstrate deductively that A was true. The deductive success of these tests resulted in an inductive affirmation of the hypothesis, which led to changes in assembly and flight readiness testing procedures. With these changes in place, the validated hypothesis was then inductively projected onto future flights, and subsequent flight success without replication of the failure validated the entire line of reasoning. Collectively, the results of each journey through the inferential triad formed the experience base.

The experience base comprised one leg of the three-legged technical rationale used by NASA and Morton Thiokol to accept each additional specific risk.[21] The first leg was comprised of the belief that the worst primary O-ring erosion possible was 0.090″, and that, at that level of erosion, the O-ring would still seal effectively. This represented a safety margin, but the quantification of this margin was entirely dependent on the second leg of the rationale, the experience base. The experience base, to date, was limited by a maximum primary O-ring erosion, measured post-flight, of 0.053″. Finally, the third leg was the belief that the problem was self-limiting once gas pressure within the joint had equalized.

Back to Minnameier's point: "induction can only establish truth relative to the current state of affairs. It can never be excluded that future evidence may challenge a currently well-established theory".[22] Sure enough, eventually a primary O-ring completely failed to seal, the measured erosion was 0.171″, and the secondary O-ring was compromised. An idiosyncratic causality was found; the primary O-ring had never been in a proper position to seal in the first place[23]. While this could be corrected by better assembly procedure and verified by increasing the pressure used to test O-ring pressurization prior to launch, it plainly represents the vulnerability of the experience base to falsification. To this point, the construction of risk, once flight experience had deviated from expectations, had become predicated on the control of specific risks identified through post flight evidence. Prior to the launch on which it occurred, reasons not yet experienced and identified, which ultimately lead to the failure of the primary O-ring to seal at all, represent the kind of "unknown events" described in average risk … an unsafe condition based upon the averaged estimated probabilities of all unknown events. After this launch, and with the idiosyncratic explanation, it becomes yet another specific risk. And this is where the normalization of deviance occurs.

The normalization of deviance occurs when the specific risks identified through experience are conflated with average risk. In other words, experience is interpreted to color in portions of the gray area beneath the probability distribution of average risk, thus affirming the footprint of average risk as the experience base develops, and so confirming the thin tails of the Gaussian bell curve as boundaries of possible conditions. This is an enormously important idea to get one's head around.

In Figure 6.4, we can see a normal, Gaussian distribution of the probability of risk. Hopefully, the system has been designed so that the really high-risk events, such as O-ring failures, are of very low probability. The low-risk events that happen fairly frequently are mostly found in your Minimum Equipment List. And, as long as the medium- and high-risk events are occurring within the expected probabilities, the system is functioning properly.

But what happens when a high-risk event occurs with greater frequency than expected? These are represented by the black dots outside of the distribution boundary in Figure 6.4. The first thing we will do is investigate and try to explain this. In the case of the O-ring failures, this is exactly the process followed. With careful

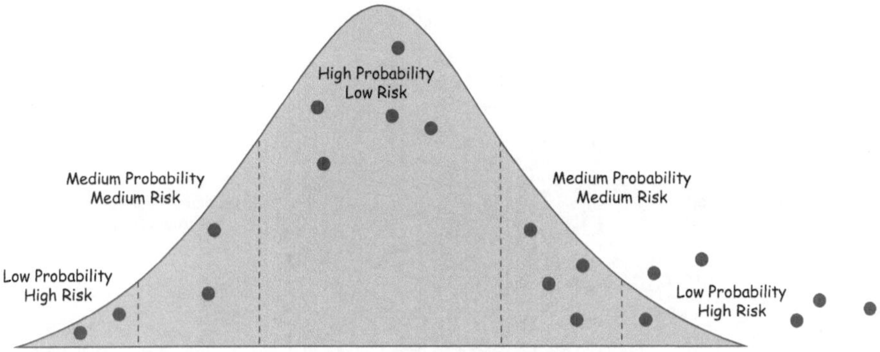

FIGURE 6.4 Risk density of an experience base.

analysis, we can explain what has happened. And, given that the vehicle has survived, we will instinctively reduce the severity of the threat assigned to that event. So, while it seems to be happening more frequently than we had expected, it isn't really as severe a threat as we had originally thought, because we understand what happened, or at least, we think we do.

Figure 6.5 shows a cartoonish view of this process. We process each one of these events, which begin to form our experience base, through the inferential triad. As we investigate and analyze the rare, higher-risk events, we develop isolated pieces of knowledge about the specific risk that these events pose. With that knowledge comes the ability to control the particular event, and thus a reduction in the threat severity it poses. The known data points, now grayed out, have migrated into specific areas of detailed knowledge. As Figures 6.4 and 6.5 show, this comprehension of small areas within the overall probabilistic risk profile creates a sense that some generalized area of the probability distribution is now controlled. The more O-ring problems we have investigated and mitigated, the more we know about the risks.

In Chapter 4, we discussed the migration of probabilistic risk in landing distance planning toward a more specific risk approach gained through quantification

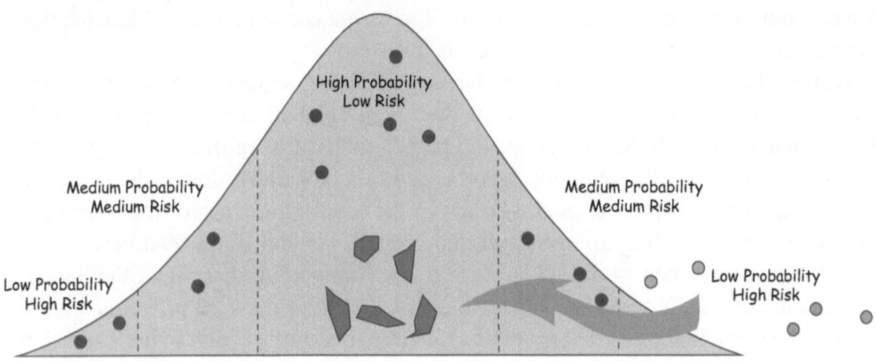

FIGURE 6.5 Migration of deviant risks.

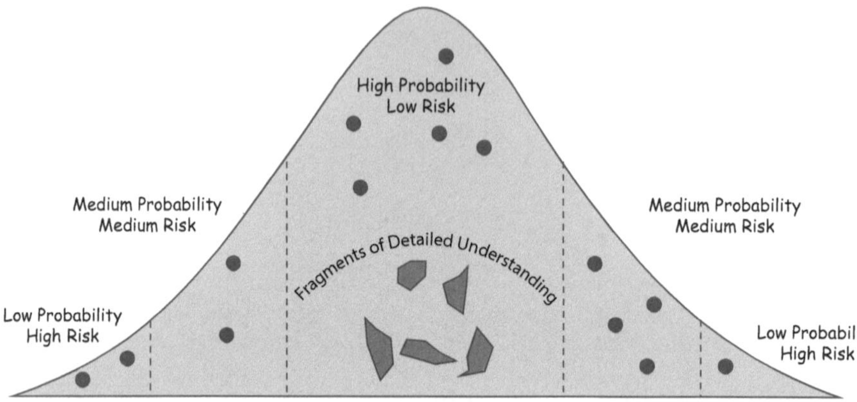

FIGURE 6.6 The conflation of specific risks into average risk.

of several variables such as approach speed additives, runway slope, etc. We also discussed the initial approach taken toward fuel tank ignitions, that of better controlling possible ignition sources through better manufacture and inspection. Figures 6.4 and 6.5 capture these processes as well. As more explanatory information is associated with specific instances of low-probability events, controls can be applied that either decrease their probability even further, or, through better understanding of the dynamics, the risk can be re-evaluated as less than originally expected, which will tolerate a higher frequency, resulting in something like that shown in Figure 6.6.

This is the natural trajectory of a belief in Dekker's naive Newtonian scientism, the idea that total knowledge of the world is achievable; the greater the extent of the experience base, the fewer unknown events there are to influence an average risk probability distribution. It is a narrative trajectory that affirms the original construction of risk, comfortably reducing cognitive dissonance, rather than drawing the more conservative conclusion that the very occurrence of each specific risk may show the probability distribution to be unstable. It is an artifact of the ampliative characteristic of abductive reasoning … more specifically, that the conclusions used in the construction of risk go beyond the limits of the original premises. Such a construction of risk ignores the possibility, and severe consequences, of falsification. Eventually, the construction of risk itself is inverted.

During the telecon on the night of January 27, 1986, some of the engineers with Morton Thiokol detected a significant paradigm shift in the approach taken by NASA management.[24] To that point, the perception had been that each flight must be proven safe to fly. Throughout the process of the normalization of deviance, the margins intended to protect against error in the assumptions used in average risk had been incrementally compromised and eroded, but the flight readiness review process remained open to the possibility of falsification. On that night, the perception changed; now, NASA was suggesting that the flight must be proven *unsafe* to fly. They essentially demanded proof of a tangible specific risk, and in the absence of such proof, they abandoned all consideration of falsification and willfully excluded the possibility that future evidence may challenge a currently well-established

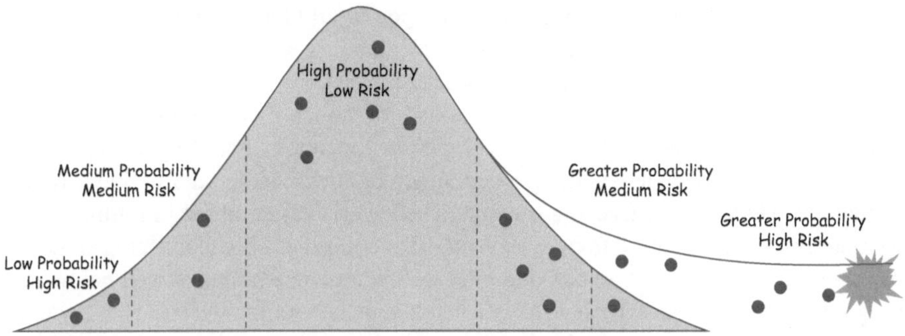

FIGURE 6.7 A fat-tailed probability distribution.

theory. At that point, the conflation of specific risk with average risk was complete. Taleb's concept of epistemic arrogance took over and closed off any possibility of acknowledging things that "disconfirm, are unpleasant, feel uncertain, seem possible, are implicit, and are contested".[25]

What the appearance of high-risk, low-probability events at a greater-than-expected frequency really means is that the originally assumed, normal distribution of probability density is wrong. The distribution has a fat tail, meaning that more events occur outside of three standard deviations from the mean than would occur in a perfect Gaussian distribution, as shown in Figure 6.7. The simple fact that some of these unexpected occurrences have been survived, and explained, and managed, does nothing to validate the original estimation of probability. It was wrong, it is still wrong, and there are quite likely more unexpected events out there. Eventually, the explanation and control of one of them will not occur until it is far too late. And we get the Challenger explosion.

On October 30, 1941, Northwest Airlines Flight 5, a DC-3, crashed on approach at Moorhead, Minnesota.[26] While conducting a "look-see" descent behind a cold front, the aircraft encountered significant icing and entered a stall from which recovery was not successful. The captain was the sole survivor of the fifteen people on board. During the investigation, he said that upon reaching an altitude of 1,500 feet, "we did start to pick up quite a lot more ice". However, the report went on to say that, "… having on previous flights experienced what he considered heavier ice accumulation, he still was not unduly concerned". The airplane stalled almost immediately after leveling at 1,500 feet.

While the captain was mindful enough to note "a lot more ice", his interpretation of his previous successful experience was that he need not be "unduly concerned". He had many hours of flying in ice, and like most pilots, he had abductively reasoned early in his experience that ice of limited quantity and limited accretion rates was not a threat. This was verified inductively by subsequent success in apparently similar ice. This, too, is the normalization of deviance. To him, it was another encounter with a defined specific risk that he no longer considered to represent an unsafe condition resulting from the average estimated probabilities of all unknown events. Deice boots do not maintain a completely clean wing (particularly if not cycled consistently), and everyone knew that you would have to put up with some ice accretion when flying in such conditions. He had migrated these high-risk events over into the area of higher frequency, lower risk events.

But the problem with ice shapes is that no two of them are ever really the same. As we discussed in Chapter 5, characteristics such as roughness, horn angle, and chordwise extent can vary widely with airspeed, angle of attack, temperature, liquid water content, droplet size, and freezing fraction. These variables combine in infinite ways to create emergent threats, "unknown events" that comprise a body of average risk, in this case the probability of an aerodynamic stall at an angle of attack well below the design stalling angle. An experienced conclusion that these unknown events fall within the scope of a known specific risk takes the word "discriminatory" right out of Weick's definition of mindfulness. Mindfulness is intended to be the brake on the open-ended ampliative nature of both abductive and inductive conclusions. If Hume originally described induction as a process which could infer that "instances of which we have no experience resemble those of which we have had experience",[27] a rich awareness of discriminatory detail is the method by which we can infer the opposite: that this instance does not resemble *any* instances of which we have had experience.

One of the simplest examples of the normalization of deviance is the case of popped circuit breakers. We've all dealt with these situations many times. It is ridiculously easy to simply reset the breaker, and if it doesn't pop again, just press on. In fact, when I started flying, most operators had a policy that allowed one reset of circuit breakers, with the single exception of fuel pump breakers. After considerable experience with electrical fires, including but not limited to arc tracking, the industry became wise to the actual fat tails of this probability distribution. Standard policy changed; no resets are allowed unless a specific QRH procedure calls for manually opening the breaker and resetting it. This can be challenging to comply with; it is typically discovered at the gate, policy says call maintenance, the maintenance guy comes out, resets the breaker, looks around, and signs it off. What they are really supposed to be doing is executing one or more specific tests and checks to ensure that the popped breaker was just that … a spurious popped breaker. As we discussed in Chapter 5, an arc tracking event will not necessarily cause the breaker to pop in the first place. If it does, it may not pop it again. You can *never* interpret your experience with popped circuit breakers to mean that the absence of a second popping event is an indication of no fault.

In their book "Managing the Unexpected", Weick and Sutcliffe explain that "participants in high reliability organizations know the limits of their own knowledge and experience. They seldom 'mistake the change of a feeling of doubt into a feeling of assurance *as knowledge*'" (Italics added).[28] Yet this is precisely what a pilot does as he or she gains experience … move from the initial feeling of doubt into an experienced feeling of assurance and then interpret that as knowledge.

Shift gears for a moment. As I said earlier, the linguistic adaptability of the phrase "normalization of deviance" opens it up to multiple interpretations. In particular, the word deviance is very, very close to another word we use frequently in the study of accidents … violations.

In the past few years, the nomenclature underlying the normalization of deviance has coalesced somewhat around what is known as violation behavior, in particular offering an explanation for routine violations. James Reason defined four categories of human error: slips, lapses, mistakes, and violations.[29] Of these, all are unintentional, with the exception of violations, which are considered intentional. A violation is a departure from a rule-based paradigm, such as a regulation or a standard operating procedure. Reason further split violations up into four subcategories: routine, optimizing, situational, and exceptional. Of these, routine violations are by far the most common.

With time, however, the concept of violation behavior has taken a turn back toward the language of the tenth axiom. Violation behavior is characterized by a specific intention to deviate from regulations, rules, or standard operation procedures. You can almost draw a straight line from this definition right back to Adam taking a bite of the apple; it is presented as a fall from grace. The characterization of individuals prone to violation behavior as wolves and more compliant individuals as sheep, as reported by Patrick Hudson in reference to work done on safety in the North Sea oil fields, further cements the notion of violation behavior as opportunistic and self-interested. Hudson describes the "sheep" as "guardians of the high standard",[30] which tends to smack of a cross between a man of the cloth and one of those kids in school who never missed an opportunity to tell the teacher when you were busily crafting spitballs in the back of the room.

Dave English and Russell Branaghan have proposed a newer taxonomy of violation behavior, which they argue to be more consistent with Human Factors Analysis and Classification System (HFACS). They point out that

> In common law, the concepts of recklessness and negligence are included inmens rea (Latin for guilty mind), so making violation behavior (actus reus) a criminal act.[31]

Their taxonomy includes the terms malevolent, indolent, and hedonic. Implicit in this concept are Frederick Taylor's ideas of a worker who is unable to see the big picture, so-to-speak, who in any case probably lacks the intellectual capacity to organize his own work efficiently, as well as Adam Smith's economic notions of self-interest as a primal motivator. Essentially, this takes the argument right back to Varney's example of weak character, in which the fellow flying across the Atlantic starts out by first buzzing the blonde at the cigar factory. There are indeed plenty of examples of reckless operation in the accident archives, and to the extent that English and Branaghan's classification addresses those cases, it is accurate and useful. But veering too far into the sphere of inherent depravity misses the core point. Pilots, like engineers, normalize deviance based on subtle misinterpretations of extensive technical experience, which is a far cry from indolence.

Vaughan describes this as a social normalization of deviance, in which people become so accustomed to deviating from the norm that they don't recognize these deviations as abnormal, despite having drifted far beyond their own rules for basic safety.[32] The key to capturing the dichotomy is to consider that you can do precisely what Vaughan has described without any motivations of indolence, hedonism, or any malevolent ill will whatsoever.

In fact, Vaughan argues that the principal problem behind the decision to launch the Challenger was a perception on everyone's part that they had complied thoroughly with all of the rules.[33] She contests the public perception, largely created by media accounts as well as the spin of the Presidential Commission, that NASA management operated on the basis of "amoral calculation" (yet another reach back to Adam Smith). In other words, rather than making a launch decision on the basis of cost-benefit trumping morality, they actually made the decision based on the perception that the engineering work had complied with all of the existing rules, requirements, and standards, thus giving them confidence that the system had functioned properly and the risk assessment was correct. There was absolutely no one on the telecon the night before the launch who could be even remotely classified according to terms such as malevolent, indolent, and hedonic.

NOTES

1. Diane Vaughan, *The Challenger Launch Decision: Risky Technology, Culture, and Deviance at NASA*, University of Chicago Press. Kindle Edition.
2. Transportation Safety Board of Canada, Aviation Investigation Report A11H0002, Controlled Flight Into Terrain, Bradley Air Services Limited (First Air) Boeing 737-210c, C-GNWN, Resolute Bay, Nunavut, August 20, 2011.
3. Leon Festinger, Henry W. Riecken, Stanley Schachter, *When Prophecy Fails*, Kindle Edition, Wilder Publications, 2014, 20.
4. Eddie Harmon-Jones, David M. Amodio, Cindy Harmon-Jones, "Action-Based Model of Dissonance: A Review, Integration, and Expansion of Conceptions of Cognitive Conflict", in Mark P. Zanna, Editor, *Advances in Experimental Social Psychology*, Vol. 41, Academic Press, 2009, 128.
5. Karl Weick, Kathleen Sutcliffe, "Managing The Unexpected", 2nd Managing the Unexpected in Wildland Prescribed Fire Workshop, Jacksonville, Florida, February 28, 2005.
6. Government of Portugal, Ministério Das Obras Públicas, Transportes E Comunicações, Gabinete De Prevenção E Investigação De Acidentes Com Aeronaves, Accident Investigation Final Report, All Engines-Out Landing due to Fuel Exhaustion, Air Transat Airbus A330-243, C-GITS, Lajes, Azores, Portugal August 24, 2001, 77.
7. Ibid., 60.
8. Ibid.
9. Transportation Safety Board of Canada, 125.
10. Ibid., 83.
11. Eddie Harmon-Jones, Cindy Harmon-Jones, "Cognitive Dissonance Theory after 50 Years of Development", *Zeitschrift für Sozialpsychologie,* Vol. 38, No. 1, 2007, 7–16.
12. Gerhard Minnameier, "The Logicality of Abduction, Deduction and Induction", in Mats Bergman, Sami Paavola, Ahti-Veikko Pietarinen, & Henrik Rydenfelt, Editors, *Ideas in Action: Proceedings of the Applying Peirce Conference.* Nordic Studies in Pragmatism 1. Nordic Pragmatism Network, 2010, 239–251.
13. Harry G. Frankfurt, "Peirce's Notion of Abduction", *The Journal of Philosophy*, Vol. 55, No. 14, July 3, 1958 https://doi.org/10.2307/2021966, 594.
14. Minnameier, 244.
15. David Hume, *Treatise of Human Nature*, T 1.3.6.4, SBN 88-9, 1739, https://davidhume.org/texts/t/1/3/6.
16. David Hume, *An Enquiry concerning Human Understanding*, Oxford University Press, 2007, 25.
17. Minnameier, 242.
18. Noel E. Boulting, Charles S. Peirce's Idea of Ultimate Reality and Meaning Related to Humanity's Ultimate Future as Seen through Scientific Inquiry, https://www.utpjournals.press/doi/pdf/10.3138/uram.16.1-2.9, March 14, 2022.

19. Karl Popper, *Conjectures and Refutations: The Growth of Scientific Knowledge*, Basic Books. Kindle Edition, 1962, 49.
20. Minnameier, 247.
21. Vaughan.
22. Minnameier, 242.
23. Ibid.
24. Ibid.
25. Karl Weick, Kathleen Sutcliffe, "Managing the Unexpected", 2nd Managing the Unexpected in Wildland Prescribed Fire Workshop, Jacksonville, Florida, February 28, 2005.
26. Civil Aeronautics Board, Docket No. SA-53, "Of the Investigation of an Accident Involving Civil Aircraft of the United States NC 21712 which Occurred Near Moorhead, Minnesota on October 30, 1941", Released January 31, 1942.
27. Hume, *Treatise of Human Nature.*
28. Karl Weick, Kathleen Sutcliffe, *Managing the Unexpected: Assuring High Performance in an Age of Complexity* Wiley. Kindle Edition, 2015, 116.
29. James Reason, "Human Factors: A Personal Perspective", Human Factors Seminar, Helsinki, Finland, February 13, 2006.
30. Patrick Hudson, "Non-Adherence to Procedures: Distinguishing Errors and Violations", 11th Human Factors Symposium, Melbourne, Australia, July 2000.
31. Dave English, Russell Branaghan, "An Empirically Derived Taxonomy of Pilot Violation Behavior", *Safety Science*, February 2012, doi: 10.1016/j.ssci.2011.08.009.
32. Interview with Diane Vaughan, Digital Cube Interactive, http://www.consultingnewsline.com/Info/Vie%20du%20Conseil/Le%20Consultant%20du%20mois/Diane%20Vaughan%20(English).html, March 20, 2022.
33. Vaughan, *The Challenger Launch Decision.*

7 Automation

I have been known, on occasion, to talk to the autopilot. "Why on earth are you closing the throttles now?", or "What? Who told you to fly at 210 knots?" It's possible that this could be a little unnerving to an unsuspecting first officer, but there are occasions when it is necessary to question the autopilot's intentions or even its situational awareness. Sometimes I have to intervene … "No, no, let's not do it that way … here, let's try this mode …" And every so often, "Oh for goodness' sake, stop making this harder than it is …" a comment usually associated with disconnecting the thing.

Some of that comes from the early days of my career, the first 5,000 hours of which involved Convairs and Metroliners with no autopilots and no flight directors. We hand flew all day, well, more precisely all night, every night. This was a pattern only gradually altered by flying the 727, whose autopilot was equipped with an input control that we commonly referred to as the "lurch lever" because the spring tensions were not well-calibrated to the G tolerances of the typical passenger's posterior. On legs under an hour, many of us never engaged the autopilot at all; nor did we activate the flight directors unless we were flying an instrument approach. We simply flew pitch and power like we always had.

But most of it comes from a strategy to manage two parallel and integrated situational awarenesses; the old, original one … where are we, where is the ground, where are we going and at what angle of attack … with a new one: where does the autoflight system think we are, where does it think we want to go, how is it going to get us there and, perhaps of separate but equal importance, where are we within one or more flight envelopes that it is designed to protect us from departing? Both situational awarenesses are vital to safety. But with the advent of the second awareness, the automation awareness, it has become common for the authorities, manufacturers, and various other august bodies of expertise to start describing pilots as "systems operators" or "system managers". This is not really a new idea. In 1953, Guy Murchie, writing in his book "Song of the Sky", rather presciently predicted

> A maplike screen on which will be projected pips of light representing not only his own position but those of other craft, enabling him to monitor the traffic situation continuously and check navigation by eyesight in the densest cloud.[1]

This is a curiously accurate description of an FMS-driven Navigation Display with TCAS superimposed. By 1959, General Pete Quesada, the first FAA Administrator, observed that, with respect to military pilots, "The day of the throttle jockey is past. He is becoming a true professional, a manager of complex weapons systems".[2]

But back in 1939, writing in his masterpiece, Wind Sand and Stars, the French airmail pilot Antoine de St. Exupery anticipated how we might lose control of this evolution. He wrote that

> In the enthusiasm of our rapid mechanical conquests we have overlooked some things. We have perhaps driven men into the service of the machine, instead of building machinery for the service of man.[3]

DOI: 10.1201/9781003369677-7

It is easy to intuit how the concept of a manager of systems veers toward a man in the service of the machine. With the acceleration of automation in the cockpit and the mishaps and accidents that have resulted, it seems to me that we have never truly resolved St. Exupery's point. On the one hand, the pilot-in-command remains the final authority as to the operation of the aircraft. On the other, the pilot is an operator of complex systems that he is no longer expected to understand.

TWO APPROACHES TO AUTOMATION: CAPABILITY AND CYBERNETIC

A few years ago, Major David Blair and Captain Nick Helms, both officers in the United States Air Force, published a thoughtful paper entitled "The Swarm, the Cloud, and the Importance of Getting There First", a treatise on remotely piloted aircraft (RPAs) operated by the Air Force. They concisely and carefully captured St. Exupery's dichotomy in more contemporary terms:

> The first truth of special operations holds that humans are more important than hardware. In other words, technology exists to enable people to fulfill the mission. This is the capabilities view of technology: machines are amplifiers of human will, better enabling them to make something of their world. By exercising dominion through technology, people gain greater command over their environment. The alternative is that humans are important to operate the hardware – that people are subsystems within larger socio-mechanical constructs. This view, cybernetics, encloses people within closed control loops that regulate systemic variables within set parameters. Rather than human versus machine, the true discussion about the future of RPAs addresses capabilities versus cybernetics.[4]

The original intent of contemporary cockpit automation arose from the capabilities view of technology, in particular the capability to optimize aerodynamic efficiency while also optimizing airspace utilization. This was, and still is, clearly a machine in the service of man. The intent of automation began to migrate toward the cybernetics view with the notion that we could automate human error out of the equation. In my experience, this migration happened about the same time we transitioned from experienced instructors' hand-drawing schematics on whiteboards to well-meaning but very inexperienced people flipping Powerpoint slides salted with schematics from the maintenance manual. Cockpit automation is today widely discussed and trained from the cybernetics view of technology. This has been powerfully reinforced by the early understanding of human factors as a deterministic, predictable discipline, indeed, by the fundamental understanding of behavior from the deterministic view of neuroscience.

In their 2014 report entitled "A Practical Guide for Improving Flight Path Monitoring", the Flight Safety Foundation noted that

> Multiple studies have shown that many pilots poorly understand aspects of autoflight modes, in part because training emphasizes correct "button pushing" over developing accurate mental models. Simply stated, it is impossible to monitor a complex system if a pilot isn't sure how to correctly operate that system or what type of aircraft

performance can be expected from each autoflight mode. A pilot who has an accurate mental model of the autoflight system can then learn how to use each mode and will be able to accurately predict what the aircraft will do next in a given mode in each specific situation.[5]

THE EVOLUTION OF AUTOMATION

A short trip through Mr. Peabody's Wayback Machine will place us in a new-hire flight engineer classroom. The instructor is a retired chief master sergeant, and he is diagramming by hand the disassembly, piece by piece, of an air conditioning pack. By the time he is done, the new pilots will thoroughly understand how a pack works and therefore have a solid grasp of what they are looking at on the pack temp gauge … at least that was the plan in those days.

In order to get rid of the flight engineer, we had to get rid of the pack temperature gauge. The thinking was that by automating the systems and improving the system status annunciations, we could make the task of monitoring systems much simpler. As we automated, we also watered down the ground school; there was no longer any reason to truly understand the system at a component level, since the automation would tell you all you needed to know. This is precisely the trajectory that Murchie had in mind when, continuing his 1953 description of a future cockpit, he said that

> Elimination of everything unessential is a big load off the crew's brains. When the flight engineer wants to check whether his battery generators are working he used to have to read a dial needle pointing to numbers of amperes of charge or discharge. In the future he will only see a green or red light indicating "yes" or "no". With fifty such indicators shorn of their wool, the crew will be spared much of the dangerous excess of information from which they have long had to select, abstract, interpolate, extrapolate, derive, and ignore – sometimes literally to the point of death. The airplane will enter a new phase of progress.[6]

But along the way, I believe a very subtle paradigm shift occurred. Back in the day, we had a vague idea of approximately where we were in space. Between the A-N ranges, ADF pointers, and LORAN systems, we were generally sure of which hemisphere we were flying in, and with some skill, we could place the airplane over a runway threshold safely and reliably, albeit with little surety of exactly where we had been in the process of getting there. Whilst sorting out the bearings, radials, and tones, it was essential to keep all one hundred and twelve cylinders lubricated, firing properly and not consuming more gasoline than was absolutely necessary. Monitoring had a great deal to do with aircraft systems and less to do with the flight path. The flight path was more a matter of technique as long as one avoided an unintended stall.

But at the same time we were automating away little dials pointing at numbers indicating amperes, we were increasing airspace occupancy exponentially. Frequency, frequency, frequency. More flights, more options, more consumer choice, more tailored load factors, more capacity, and then more capacity management … all while still operating approximately the same number of outer markers as we have for over sixty years. Capacity is choked; this leads immediately to tightening the longitudinal and vertical spacing between aircraft, as well as such things as PBN,

RVSM, RNAV departures and arrivals, and the like. All of this is basically intended to obtain the maximum arrival rate possible for each runway at each terminal.

So, the importance of flight path management has become supreme and highly automated. The improvements in flight path management began as a capability approach to automation, an amplifier of human will, gaining greater command over a limited environment. But the work of designing an optimized airspace infrastructure, coupled with a cold statistical analysis of the threat posed to public safety by capacity maximization, created a need for objective, uniform, and repeatable degree of predictability in human performance. At the same time, the development of automation, and particularly binary decision gates, gave rise to a model of human cognitive function that was, essentially, binary. Fortunately, the boffins were in possession of Herbert Simon's old idea that "Two persons, given the same skills, the same objectives and values, the same knowledge and information, can rationally decide only upon the same course of action".[7] Viewed from this perspective, enclosing "people within closed control loops that regulate systemic variables within set parameters"[8] makes perfect sense … the automation paradigm shifted from the capability approach to the cybernetic approach.

In this manner, the airspace infrastructure has evolved into the kind of larger socio-mechanical construct that Blair and Helms described, in which people are subsystems. Along the way, the shift in paradigm, as well as a culture mesmerized with automation and digitization, slowly, and probably unwittingly, displaced procedural knowledge with declarative knowledge.

Simon Wood, of Cranfield University, has described declarative knowledge as "the knowledge that the system works in a certain way", and contrasted this with procedural knowledge, which he describes as "knowing how to use the system in context". He explains that

> The basic skills associated with "manually flying" an aircraft are predominantly based on procedural knowledge, i.e. how to achieve the task. However, the use of automation to control the flight path of an aircraft is taught as declarative knowledge. Pilots are required to manage systems based on a knowledge that the autoflight system works in a particular fashion. So, the pilot is faced with the same operational task of controlling the flight path but employs two different strategies of cognitive behaviour depending upon whether the task is manually or automatically executed.[9]

It is important to stop for a minute and put this concept under a microscope. In the days of the flight engineer, declarative knowledge and procedural knowledge were more or less balanced, and they were integrated. Declarative knowledge supported procedural knowledge and we were taught both. If you wanted to get the generator on line, you were going to have to synch the generator frequency to the bus frequency; you had to understand how this worked, and you had to be able to make it work, because it wasn't going to do it by itself.

But right there, at that inflection point, is where the problems of automation gain a foothold, precisely because automated systems *will* do it by themselves. It is no longer a matter of procedurally operating a system; it is a matter of watching the system procedurally operate itself. When the Flight Safety Foundation describes an "accurate mental model which will enable the pilot to predict what the airplane will do next in a given mode for each specific situation",[10] they are referring entirely to declarative knowledge, a knowledge of how the system works, with the expectation that the pilot's speed of cognition will exceed the system's own procedural operation. In the old days, the pilot's speed of cognition controlled the procedural operation. Nothing would happen until you were ready for it to happen, because you had to make it happen. You could get behind the airplane moving in space and you could get behind the situation in time, but it was pretty hard to get behind the systems. Today, you'd better be on your toes, because the automated system is going, with or without you. Indeed, the very phrase "predict what the airplane will do next", as if this were a matter of conjecture, implies that the airplane has a mind of its own.

Yet the premise behind watered-down training is that the modern, sophisticated fly-by-wire airplane is too complicated for the pilot to fully understand, and thus he or she has no need for extensive knowledge of the aircraft design and architecture. This is entirely in line with Murchie's 1953 prediction that the crew "be spared much of the dangerous excess of information from which they have long had to select, abstract, interpolate, extrapolate, derive, and ignore".[11] Sixty years later, in the 2013 report, "Operational Use of Flight Management Systems", the Performance Based Operations Aviation Rulemaking Committee said that

> Pilot knowledge of the basic airplane systems is not as detailed as in the past. The WG recognizes that in the past, information was trained that was not needed or beneficial. The concern is that depth of systems knowledge may now be insufficient, and this may be operator dependent.[12]

And so we arrive at the rather matter-of-fact condescension expressed in a pivotal statement following the 737 Max debacle, in which a Boeing official stated to the Wall Street Journal that the company had decided not to provide details regarding the Maneuver Characteristics Augmentation System (MCAS) because they were afraid that such information was more than average pilots needed or could digest.[13]

THE PATH TO ETHIOPIAN 302

On March 10, 2019, Ethiopian 302, a Boeing 737 Max, departed Addis Ababa enroute to Nairobi. Within seconds after liftoff, a faulty angle of attack signal was generated and the stick shaker activated. The left flight director command bars deviated significantly from the right. The master caution light illuminated. Over the course of the next ten seconds, the crew attempted to engage the autopilot twice. Approximately fifteen seconds later, they successfully engaged the autopilot. After a further twenty seconds, the captain called for the flaps to be retracted. At about the same time the flaps completed retraction, the autopilot again disengaged. This completed the necessary conditions to activate the MCAS, and pitch control was critically compromised. In the process, the autothrottle was never disconnected, the power remained in the N1 mode at takeoff thrust, and since the pitch attitude was severely depressed, the airspeed went right through the barber pole, driving the aerodynamic loads on the stabilizer well beyond the ability of the manual trim wheel to manage.[14]

As Edgar Morin had predicted, "As soon as a person begins any action whatsoever, the action starts to escape from his intentions. It enters into a sphere of interactions and is finally grasped by the environment in a way that may be contrary to the initial intention".[15] Pretty much everything the Ethiopian crew did escaped their intentions. There are few accidents in the history of commercial aviation that make this point more painfully.

The path that the industry has followed to get to the Ethiopian crash is embedded within the path followed to maximize capacity within the airspace infrastructure and the paradigm shift from a capability view of automation to a cybernetic view. The path is delineated by two features: the proceduralization of safety and the concept of the "ultra-safe" approach to risk.

Herbert Simon's idea that two persons can rationally only decide upon the same course of action is really nothing more than the automation of human behavior. Standard operating procedures, when interpreted in this way, become lines of code. People schooled in automation easily interpret the human being as a machine. While the school of human factors has evolved toward understanding the complexity and emergent characteristics of human behavior, the school of automation, which is hopelessly immersed in top-down control, embarked on the Simonesque mission to preset the variables to fail-safe values.

Isabelle Fucks and Yves Dien have described the paradox of procedures this way:

> There are two ordinary potential uses of procedures: either procedure is an aid, a reminder; or a rule, a standard to comply with, to enforce. In one case, the user's skills and competences are acknowledged; she has some room for maneuver and some responsibilities. In the other case, the user has no responsibility, she is controlled by the procedure.[16]

As we discussed in Chapter 1, procedures are easily manufactured and manipulated as a means of exercising control over individual behaviors and performance. Mathilde Bourrier and Corinne Bieder explained it this way:

> Moreover, proceduralization is a privileged axis through which bureaucratization operates. Its history is attached to the numerous attempts at tracing, controlling and anticipating social activities. Proceduralization captures two sides that are totally embedded: control on the one hand and explicitation on the other hand. It is no mystery that written procedures not only indicate how tasks should be performed but they also offer the possibility for management and regulators to control workers' compliance.[17]

Controlling and anticipating social activities, and controlling workers' compliance, naturally generate a form of pushback from sentient human beings who desire to create an identity of craftsmanship and responsibility. Because such bureaucratization inevitably falls well short of what actual operations require, it also generates shortcuts and bridges, essentially procedural modifications on the fly, to get things done. This is the world of routine and optimizing violations.

And this is the original point behind Rene Amalberti's idea of an initial space of safe operation. He described the initial space of safe operation as designed around rules and procedures, including various constraints and fail-safe procedures, all of which are intended to serve as defenses against human error and constrain possible human actions.[18] Amalberti was primarily concerned with constraining the drift away from the centerline of that space, a drift which is continuously influenced by broader social and organizational forces.[19] His argument is that the system will naturally settle into a form of stability some distance away from the centerline of that initial space of safe operation, but still within the "borderline tolerated conditions of use"; however, the margin will be degraded. To put it in another hypothetical sense, we are taught to land on the centerline of the runway, but eventually the pressure to improve efficiency and productivity leads us to regularly land off the centerline, on the side of the runway closest to the terminal so that we can minimize taxi time, but still with a dozen feet or so between the main gear and the edge of the pavement. Safe enough?

Amalberti's description of rules, procedures, and constraints designed to defend against human error is the definition of standard operating procedures. In this construction, the initial space of safe operation is created by the standard operating procedure. The goal is to remain within that space. Any drift away from the centerline of the space, through things like the normalization of deviance, misinterpretation of experience, or misunderstanding of risk, erodes the margins. The margins give us time and space to spot an action that has departed our intentions and shoot it down.

But right here is where we arrive at a critical bifurcation point in the construction of safety. As Fucks and Dien have said, there are two uses of procedures; either as an aid, which recognizes skill and competence, allowing room for maneuver and assigning responsibility, or as a rule, which negates local responsibility and locates the pilot as a subsystem within the larger socio-technical construct. Note the very clear parallel between these two uses of procedures and the capability view – cybernetic view dichotomy described by Blair and Helms.

Amalberti located commercial aviation within what he describes as the "ultra-safe" model of risk, which he describes this way:

> With this approach we turn radically away from reliance on human skills and ingenuity towards a reliance on standardization, automation and avoidance of risk wherever possible. Professionalism in these contexts still requires very high levels of skill but the skills consists primarily in the execution of known and practiced routines, covering both routine operations and emergencies. Ideally, there is no need to rely on exceptional expertise even when dealing with emergencies, such as an engine fire on a commercial aircraft. Instead comprehensive preparation and training allows all operators to meet the required standards of performance and be skilled to the point that they are interchangeable within their respective roles.
>
> This approach relies heavily on external oversight and the control of hazards which makes it possible to avoid situations in which frontline staff are exposed to exceptional risks. By limiting the exposure to a finite list of breakdowns and difficult situations, the industry can become almost completely procedural, both when working under normal conditions and under more difficult conditions.[20]

Note the phrase, "By limiting the exposure to a finite list of breakdowns and difficult situations ...". This is nothing more than a list of specific risks ... known-knowns. The argument that "there is no need to rely on exceptional expertise even when dealing with emergencies" is, in fact, a restatement of the cybernetic approach to automation described by Blair and Helms.

St. Exupery would have strongly disagreed with this view. He wrote, also in Wind, Sand and Stars, that

> The machine which at first blush seems a means of isolating man from the great problems of nature, actually plunges him more deeply into them. As for the peasant so for the pilot, dawn and twilight become events of consequence. His essential problems are set him by the mountain, the sea, the wind. Alone before the vast tribunal of the tempestuous sky, the pilot defends his mails and debates on terms of equality with those three elemental divinities.[21]

For the cybernetic approach to work or the ultra-safe model of risk to be valid, the operating environment must be a closed system. St. Exupery's description of a tempestuous sky is no such thing.

Edgar Morin elaborated on his idea of the ecology of action by distinguishing between strategy and program. He said that

> … the answer to the uncertainties of action is thoughtful decision, awareness of the wager, and elaboration of a strategy which allows for the complexities inherent in its finalities; can be modified in the course of action in response to chance, information, change of context; and provides for eventual torpedoing of an action that may have taken a harmful course.[22]

A thoughtful decision is nothing less than prudence, a command decision arrived at by taking counsel and judging of what you have learned. By wager he means risk, and awareness of the wager is awareness of the risk. As I said in Chapter 3, the real foot-stomper in St. Thomas' discussion of prudence is his elaboration on caution, in which he differentiates between threats of frequent occurrence that can be grasped by reason, which describes specific risk, and those "that occur rarely and by chance, and these, since they are infinite in number, cannot be grasped by reason",[23] describing probabilistic risk. In that sense, an awareness of wager is really part of the prudence that leads to a thoughtful decision.

Morin unpacked his concept of strategy further, when he said that

> Strategy should prevail over program. A program sets up a sequence of actions to be executed without variation in a stable environment, but as soon as the outside conditions are modified, the program gets stuck. Whereas strategy elaborates a scenario of action based on an appraisal of the certainties and uncertainties, the probabilities and improbabilities of the situation. The scenario may and must be modified according to information gathered along the way and hazards, mishaps or good fortune encountered. We can use short term program sequences within our strategies. But for things done in an unstable, uncertain environment, strategy imposes.[24]

The statement, "The scenario may and must be modified according to information gathered along the way and hazards, mishaps or good fortune encountered", sounds an awful lot like St. Thomas' description of circumspection. And woven into the idea of strategy prevailing over program is Thomas' insistence that foresight be the principal feature of prudence.

But Morin has gone farther than Thomas could. He differentiates between strategy and program. In St. Thomas' day, the concept of program did not exist. There were no standard procedures, algorithms, or code sequences; there was, after all, very little writing and really no printing. What Morin is saying is that we must never conflate programs … standard procedures, litanies, flows, checklists, and automated sequences … with strategy. The programs are tools with which we execute a strategy; they are not strategy in themselves.

Engaging the autopilot with the stick shaker banging away is a clear example of attempting to use a short-term program sequence and finding that program is stuck, as evidenced by the repeated attempts to engage it. Retracting the flaps on schedule is likewise a part of the same program. This was the standard operating procedure following takeoff, but their actions had been fully grasped by the environment in a way that was undoubtedly contrary to the initial intention. The environment had become highly modified and unstable. The failure to disconnect the autothrottle is a decisive indication that the program, whatever it was, was thoroughly stuck. There was no strategy to prevail over anything.

In approaching the system as a closed system, we create a systems operator who is unprepared to debate on terms of equality with the mountain, the sea, and the wind, or for that matter, with the central processing unit of the flight control computer. From the standpoint of the ultra-safe model of risk, there is no such debate. Instead, the systems operator's terms have been dictated by the set parameters within a closed control loop, designed to trigger Morin's "sequence of actions to be executed without variation in a stable environment". The foresight, the "one [that] must be the principal to which all the others are subordinate", is pre-programmed, trapped within the closed control loops, and limited to a narrow set of anticipated threats or specific risks. In some sense, Amalberti's constraints and fail-safe procedures[25] are perceived to be more or less omnipotent. This is antithetical to airmanship, because those parameters will eventually fall out of equality with St. Exupery's vast tribunal of a tempestuous sky, or today, a tempestuous automation system. The fundamental flaw in attempts to adapt the cybernetic view of technology to the problems of flight lies in the belief that we have expanded our knowledge to a point at which we have absolute, predictable, and repeatable control within a tempestuous sky, which takes us right back to Sidney Dekker's naïve Newtonian scientism. We don't have that control and likely never will. An analog world will simply swat away a digital mindset.

The ultra-safe model of risk has opted for the latter use of procedure in Fucks and Dien's explanation. The problem, of course, is that in pretty much all cultures, the pilot-in-command is assigned the final responsibility for the safety of the passengers and aircraft. In the moral philosophy of public safety, there is no place for an interpretation of procedure that obviates responsibility, a procedure that literally assigns the pilot a role as a subsystem, as if he or she were a mere component in a cybernetic design.

The cybernetic view of technology may, at first blush, seem a means of isolating the pilot from the essential problems of flight; it is easy to interpret envelope protection features this way. But at the same time, the capability view of technology still exists and still amplifies human will, better enabling us to make something of our world. By exercising dominion through technology, we gain greater command over our environment … and thus we are plunged more deeply into those essential problems. For example, we can safely increase capacity while plunging more deeply into the problems of air traffic separation (the Strategic Lateral Offset Program being one output of this particular deeper plunge). We can safely operate in lower visibility while plunging more deeply into the problems of visual aids for landing and takeoff (e.g., reduced approach and centerline lighting requirements for CAT II approaches executed with autoland, or a heads-up display). We can safely operate with optimized instability while plunging more deeply into the problems of stability (as reflected by the relaxation of Part 25 stability requirements).

The deeper plunge into the essential problems of flight brings us, inevitably, to the problem of airmanship in an automated cockpit. When St. Exupery refers to the terms of equality on which we debate those three elemental divinities, he is referring specifically to the airmanship of his day. He began his approach to this question with an understanding of the mountains, the seas, and the winds … the things which influence the sky, the great problems of nature into which the airman will shortly be plunged. He was interested in "all that happened in the sky", things which signaled "the oncoming snow, the threat of fog, or the peace of a blessed night".[26] We are still very interested in the threat of fog or oncoming snow. We are also very interested in windshear, convective available potential energy, lifted indexes, microbursts, outflow boundaries, ice crystal icing, collision coalescence freezing drizzle formation, and certainly turbulence, including mountain waves … pretty much anything that can ruin the peace of a blessed night.

To this we must add an understanding of the machine, an intuitive sense of its balance, its harmony, and its energy, a feel for how the machine leverages its precipitous position in the sky to resolve the problems of nature. To St. Exupery, the machine was the engine and flight controls all connected by stringers and spars and cables; today, we must include the complement of automation as part of the machine. For example, we must be constantly aware of pitch, power, and vertical speed, while we also scrutinize actual navigation performance (ANP) exactly as St. Exupery scrutinized the howl of the wind in the wires of his Breguet 14.

Amalberti recognized this paradox. He subsequently said that, in fact,

> Human factors are first interested in the role and position given to the end user in the system. Human factors cannot be restricted to a list of local ergonomics principles, such as shapes, colors or information flow; it first refers to a philosophy of human-machine coupling, giving the central role to Human. For example, suppose a system in which crew should be a simple executor, with a perfect instrument panel allowing a total comprehension, but without any possibility to deviate from orders, and with a permanent and close electronic cocoon overriding crew decisions. Such a system should no longer be considered as corresponding to a satisfactory human factors design. The crew cannot have a second role because the members continue to be legally and psychologically responsible for the flight. They must keep the final authority for the system.[27]

He goes on, in later work, to describe two approaches to training. The normative approach to training is precisely the cybernetic approach to automation, aimed at the ultra-safe model of risk, which Amalberti describes as in this way:

- Aviation operations can be entirely specified through standardized procedures, programs, schedules, rules, nominal tasks, certification, selection, norms.

- Safety improvement will result from more specification and more discipline from the operators.
- The human operator is one more "black-box" coupled through inputs (perceptive data) which are transformed into outputs (actions) according to specified targets (goals) using adequate transfer functions (procedures, skills).[28]

Below the pilot is one layer including normal procedures, threat and error management, and non-normal procedures. This layer interacts bi-directionally with the pilot. Figure 7.1 shows a diagram of how this approach is structured.

On the other hand, he describes the ecological approach to training with these ideas:

- Aviation operations cannot be entirely specified through standardized procedures, programs, and the like. One reason is it includes humans.
- Safety improvement will result from a better respect of the "ecology" of the system and a better acknowledgment of its self-protection mechanisms.
- Human operators are auto-organized structures, coupled through recursive processes of self-regulation, and ultimately governed by their internal intentions.[29]

Below the pilot is a layer consisting of awareness of risk and the formation of strategy. Below this layer is a layer consisting of regulations, policies,

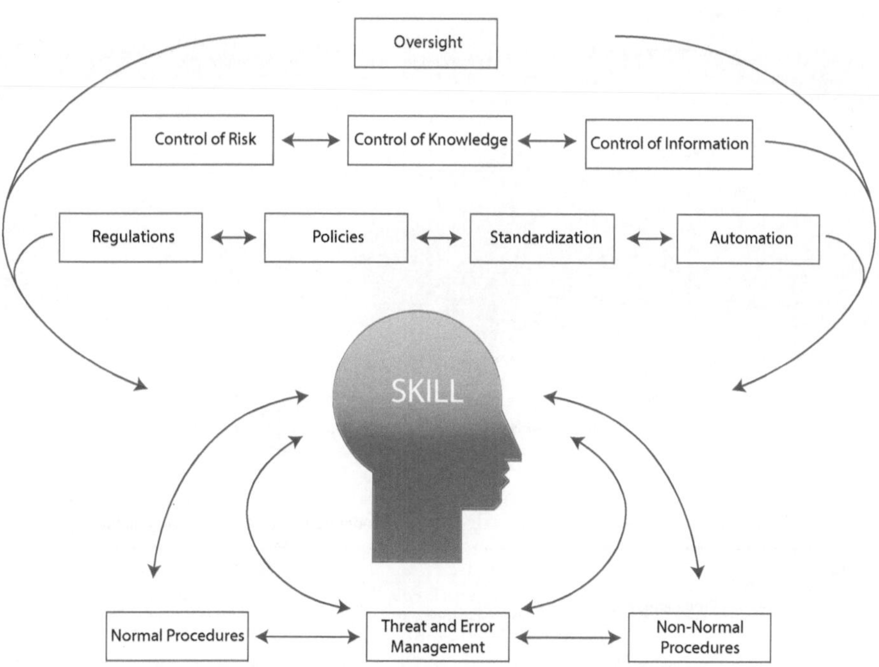

FIGURE 7.1 The cybernetic, ultra-safe model of the pilot's role.

standardization, and automation. Below this layer is a bottom layer of normal procedures, threat and error management, and non-normal procedures. All three of these layers interact bi-directionally with the pilot. Figure 7.2 shows a diagram similar in form to that of Figure 7.1, however, with these three interactive layers subordinated below the pilot to reflect human auto-organization and self-regulation.

The ecological approach to training is clearly matched to Edgar Morin's thinking about the ecology of action. Concepts such as "aviation operations cannot be entirely specified through standardized procedures" and a "better respect of the 'ecology' of the system" lay a foundation for the necessity of strategy to prevail over program. Human operators as "auto-organized" really means "self-organized", a fundamental feature of a complex system.

In St. Exupery's day, the idea of the pilot as a systems manager was unheard of, as was the contemporary suite of management school lexicon used to describe the systems manager. Terms such as discipline, professionalism, team skills, self-improvement, and skill acquisition were barely yet in anyone's vocabulary. Nor were the now-classical superlatives, such as uncompromising, optimal, systematic, and exceptional. Recent definitions of airmanship tend to include some or all of these terms; yet, in my opinion, all of them really beg the question. So what is airmanship really and how does it work in an automated cockpit?

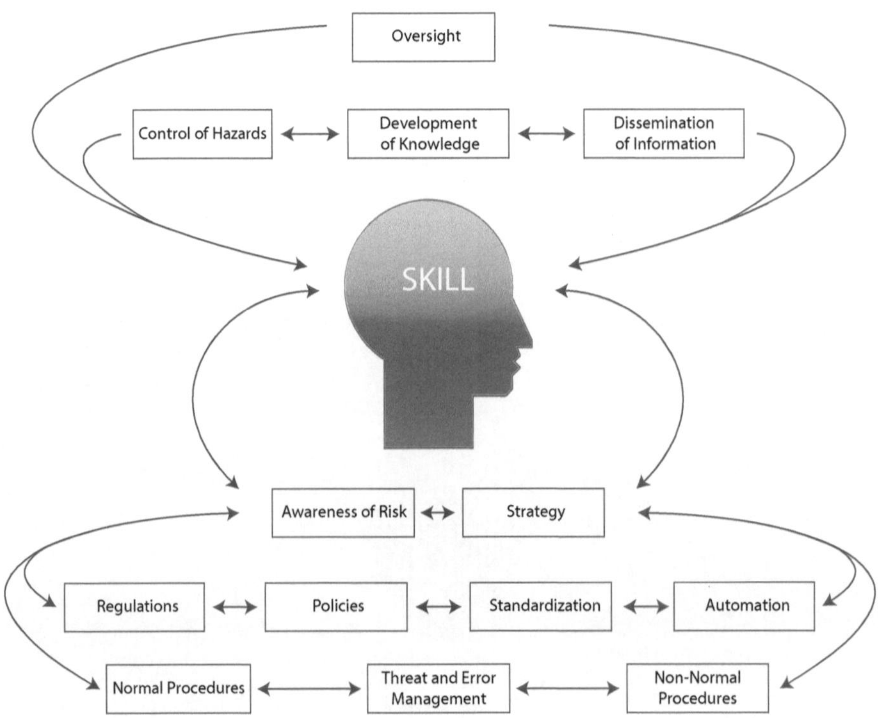

FIGURE 7.2 The ecological, capability model of the pilot's role.

A DEFINITION OF AIRMANSHIP

Let's leave the management school semantics and centuries-old conceptual structures about discipline, obedience, and compliance behind for a while. All of these are tools we use to achieve the goal; they are not the goal. Rather, let's begin by revisiting the words of FAR 91.1065(d):

> For the purpose of this subpart, competent performance of a procedure or maneuver by a person to be used as a pilot requires that the pilot be the obvious master of the aircraft, with the successful outcome of the maneuver never in doubt.[30]

The pilot, as the obvious master of the aircraft, forms the anchor of a definition of airmanship. This clearly refers to St. Exupery's idea of the machine in the service of man. It also focuses responsibility and authority for the operation of the aircraft solely with the pilot, while placing distinct emphasis on knowledge and expertise. And yet, we have to be careful of the subsequent language. The phrase "never in doubt" sounds an awful lot like the phrase "will never allow an airplane to crack up" from the tenth axiom. Both carry an implicit assumption of certainty and that flies in the face of everything we have been talking about.

Looking back through early revisions and amendments to this regulatory language, it seems likely that the elimination of uncertainty was never really the intent; the language is always qualified with the words, "The applicant's performance will be evaluated on the basis of judgment, knowledge, smoothness, and accuracy".[31] Indeed, the presence of the word judgment belies certainty. The paradox is that while we must have some degree of certainty that the flight will be successful ... if it we didn't, we would never fly ... we must incorporate uncertainty into everything we do as airmen. While we cannot accept unmitigated specific risk ... there is a microburst *alert* over the runway, so let's go somewhere else ... we have to be prepared to accept, and manage, the uncertainty associated with probabilistic risk ... there is a microburst *advisory* at the airport for our time of arrival. We have to be prepared to accept, and manage, the uncertainty associated with complex systems. We have to prevent the normalization of deviance from creating certainty where it has no right to exist. We have to be able to recognize the threat of cognitive dissonance, which occurs when certainty is unraveled. We have to spend less time on things that confirm and feel comfortable and more time concentrating on things that disconfirm and feel uncomfortable.

From this perspective, airmanship may be less about managing systems and quite a bit more about managing uncertainty. To some extent, this permeates our early flight training; we are warned by our mentors to "always have an out", and we spent a lot of time looking for good fields to use in the event of a forced landing. As young pilots, we are impressionable and can easily envision a myriad of things going wrong, and as we strive to blend into the level of competence that we believe surrounds us, we prepare as thoroughly as we can. But as we develop an experience base, certainty seems more accessible. Indeed, one of the significant problems of modern aviation is that serious failures occur extremely rarely, and the uncertainty of our early flying days is replaced with an almost inevitable, and comfortable, complacence. We get considerably less practice with resilience than we did in St. Exupery's day.

If we cannot be certain that the outcome of the maneuver, or of the flight, is "never" in doubt, we can aim toward ensuring that the outcome is never *manifestly* in doubt, meaning never exposed to an erosion of the margins. We manage uncertainty by protecting the margins; if a critical margin disappears or is compromised in any way, we have no time nor space within which to safely navigate the unruly environment, Morin's sphere of interactions, and the ecology of action can rule the day.

At this point, we can perhaps suggest a general definition of airmanship:

Airmanship is the application of both prudence and mindfulness so as to always remain the obvious master of the aircraft, and to construct, modify and execute the necessary strategies to ensure that the safe outcome of the flight is never manifestly in doubt, while always protecting the margins in anticipation of uncertainty.

While the industry dithers with the debate between the cybernetic approach to automation and the capability approach, and with the debate between Amalberti's normative and ecological approaches to training, we retain the full responsibility for the safety of the passengers and the aircraft. We must understand the margins, study how to develop strategies through the practice of prudence, and study mindfulness carefully. But at no point can the automation become the master of the aircraft.

This is the essence of the conversation I have with the autopilot.

NOTES

1. Guy Murchie, *Song of the Sky,* Houghton Mifflin, 1954, 341.
2. Major David J. Blair, Captain Nick Helms, "The Swarm, the Cloud, and the Importance of Getting There First", *Air and Space Power Journal*, July–August 2013, 33–52.
3. Antoine de St. Exupery, *Wind, Sand and Stars,* Reynal and Hitchcock, 1939, 71.
4. Blair and Helms.
5. Flight Safety Foundation, "A Practical Guide for Improving Flight Path Monitoring, Final Report of the Active Pilot Monitoring Working Group", November, 2014.
6. Murchie, 342.
7. Herbert Simon, "The Proverbs of Administration", *Public Administration Review*, Vol. VI, No. 1, Winter 1946, 64.
8. Blair and Helms.
9. United Kingdom Civil Aviation Authority, Simon Wood, CAA Paper 2004/10, "Flight Crew Reliance on Automation", December 2004.
10. Flight Safety Foundation, "A Practical Guide for Improving Flight Path Monitoring".
11. Murchie, 342.
12. Federal Aviation Administration, "Operation Use of Flight Path Management Systems, Final Report of the Performance Based Operations Aviation Rulemaking Committee, Commercial Aviation Safety Team, Flight Deck Automation Working Group", September, 2013.
13. Andy Pasztor, Andrew Tangiel, "Boeing Withheld Information on 737 Model, according to Safety Experts and Others", *Wall Street Journal*, November 13, 2018.
14. Federal Democratic Republic of Ethiopia Ministry of Transport Aircraft Accident Investigation Bureau, Report No. AI-01/19, Aircraft Accident Investigation Preliminary Report, Ethiopian Airlines Group B737-8 (MAX) Registered ET-AVJ, 28 NM South East of Addis Ababa, Bole International Airport, March 10, 2019.
15. Edgar Morin, *Seven Complex Lessons in Education for the Future*, United Nations Cultural Organization, November 1999, 45.
16. Corinne Bieder, Mathilde Bourrier, *Trapping Safety into Rules: How Desirable or Avoidable is Proceduralization*? Ashgate Publishing Ltd. Kindle Edition, Chapter 2.
17. Ibid., Introduction.

18. Rene Amalberti et al., "Violations and Migrations in Health Care: A Framework for Understanding and Management", Quality and Safety in Health Care, December 2006, doi: 10.1136/qshc.2005.015982.
19. Ibid.
20. Charles Vincent, Rene Amalberti, *Safer Healthcare, Strategies for the Real World*, Springer Open, 2016, doi: 10.1007/978-3-319-25559-0.
21. St. Exupery, 34.
22. Morin, 47.
23. Thomas Aquinas, *Summa Theologica*, translated by Fathers of the English Dominican Province, e-artnow Kindle Edition, 2013, IIa-IIae, q.49, a.8, ad.obj.3.
24. Morin, 47.
25. Rene Amalberti et al., "Violations and Migrations in Health Care: A Framework for Understanding and Management".
26. St. Exupery.
27. Rene Amalberti, "Automation in Aviation: A Human Factors Perspective", in Daniel Garland, John Wise & David Hopkin, Editors, *Aviation Human Factors*, Lawrence Erlbaum Associates, 1998, 173–192.
28. Rene Amalberti, Training Situation Assessment and Decision Making - A Rule Making Approach, Powerpoint presentation, https://www.slideserve.com/elkan/amalberti, accessed March 14, 2022.
29. Ibid.
30. United States Code of Federal Regulations, Title 14, Subchapter F, Part 91.
31. 14 CFR 61.23, Federal Register on May 2, 1962 (27 F.R, 4175).

8 Preoccupation with Failure

A pair of droning R-2800s swinging Ham Standard propellers at 900 RPM will put you to sleep in a matter of minutes, particularly at 0300 over the Great Lakes at 6,000 feet. It was pretty common for the non-flying pilot in the freighter to put his head back and doze. My approach to this was to shift my David Clarke headphones to my forehead and keep listening to ATC; this would generally keep me from truly falling asleep. And while dozing, I could monitor the health of the engines.

Back in the day, big radials were often equipped with ignition analyzers and the flight engineer, if you had one, would be adept at reading the oscilloscope output. We obviously did not have a flight engineer, and we certainly did not have ignition analyzers … we were lucky to have a working weather radar on any given night … but I learned that I could detect at least some precursors to engine misbehavior by laying my head back against the top of the seat pan and closing my eyes. Listening to the monotone of ATC transmissions while absorbing the aroma of worn leather, baking avionics, and cigarette smoke suitably heated by air from the augmentor ducts, I would wait for small vibrations.

And, sure enough, every once in a while, I would detect one. Just a tiny little rumble, felt through my neck against the seat. Once I was sure I was feeling this vibration, I would open one eye and focus on the torquemeters … specifically, the little vernier needles in the top half of the dial. And then I would wait some more. And there it was. Just a twitch, not much, but enough movement on the vernier dial to recognize. A few more cycles and I would be confident that every time I felt that little rumble in my neck, that torquemeter would twitch. And I knew, from rather more engine failures than I had signed up for, that sometime in the next couple of hours, that engine was probably going to blow a jug.

Mindfulness, coupled with an awareness of persistent vulnerability, establishes a basis for the preoccupation with failure. In his study of HROs, Weick ranks the preoccupation with failure as the first of five principles, and he describes it as encompassing three processes:

> First, they work hard to detect small, emerging failures because these may be a clue to additional failures elsewhere in the system. Second, HRO's work hard to anticipate and specify significant mistakes that they don't want to make. Third, HRO's know that people's knowledge of the system, the environment and their own group is incomplete. When people look for failures, they acknowledge the existence of incomplete knowledge.[1]

Within the first process, the key word is emerging. This does not refer to past history; rather, it refers to developing events, events that may well be unfolding on a trajectory that is surprisingly difficult to predict. These small failures are often very weak signals; little vibrations that correspond with tiny twitches of the torquemeter. They are anomalies, tiny peripeteia, little departures from the canon, or normal flow of things. The point of this first process in Weick's notion of preoccupation with

DOI: 10.1201/9781003369677-8

failure is to place the anomaly under a magnifying glass, to highlight it, to elevate the contrast between it and surrounding detail. This is an affirmative countermeasure to the idea that you cannot think of a thought until you think of it. By its definition, the notion that you cannot think of a thought until you think of it can never be completely short-circuited, but working hard to detect small, emerging failures can recapture a great deal of real estate from the forces of the unexpected.

A corollary to the idea of working hard at detecting small, emerging failures is the contemporary notion of active monitoring.

In the last couple of decades, increasing emphasis has been placed on the idea of effective monitoring skills and how to create these skills within the cadre of professional pilots. There is a wealth of academic knowledge out there about what to monitor, how to monitor, who should monitor, and the requisite skills and characteristics inherent in effective monitoring. Inevitably, the issue of discipline comes up, as effective monitoring cannot occur without some modicum of self-discipline. To a very significant extent, the failure to monitor flight path, systems, and the situation in general can be ascribed to the effectiveness of aircraft automation, along with the effectiveness of the airspace infrastructure in utilizing that automation. Inherent in these discussions are a vast array of terms that emanate from the techno-speak of system-based management theory, such as institute, instill, predictable, policies, procedures, practices, incorporate, and explicit or implicit distinctions. None of these words really has any practical value to the operating pilot. So what does?

Monitoring is an essential component of a preoccupation with failure. Go back to what Karl Weick says about mindfulness: it is, he says, a "rich awareness of discriminatory detail".[2] Being mindful means paying attention in a different way; it is to see more clearly, not to think harder and longer. You stop concentrating on those things that "confirm your hunches, are pleasant, feel certain, seem factual, are explicit, and that others agree on".[3] You start concentrating on things that "disconfirm, are unpleasant, feel uncertain, seem possible, are implicit, and are contested".[4]

I would argue that monitoring is far easier and more effective when you really understand what you are looking at. The advent of digitization has introduced another new aspect to the task of monitoring: an overwhelming amount of information easily displayed but far less easily monitored. Modern cockpits display enormous amounts of information, and they do so in not necessarily the best formats. For example, when the 767/757 type was introduced, Boeing conducted extensive research into the functionality of different formats for the novel CRT displays. They struggled with the balance between what needed to be displayed, what could be displayed, and in what format. They noted that

> A key feature of the EICAS design involved automatic color changes on gauges to denote out-of-limit engine conditions. In the midst of such a colorful array, however, such changes tended to go unnoticed.[5]

As a result of this work, Boeing drew two conclusions that are of particular relevance:

- Digital-only displays are not as good as analog-plus-digital displays, but they are not as bad as they are popularly perceived. They provide at least as good a rate indication as do analog counterparts, and their much greater precision makes them considerably better for operating close to limits. They require less space than almost any other type of display. Their major shortcoming lies in their lack of a qualitative presentation, that is, "check reading" capability.

- The seven-color CRT display is undoubtedly a more powerful medium for presentation of information than the electromechanical gauge. The electromechanical instruments found in most commercial jet airplanes now in service are, however, remarkably good displays. We seem to have forgotten how they got that way, at least in terms of exactly what their virtues are and how their features are used, individually and collectively, by the aircrew. Translating such displays into electronic "equivalents" is a deceptively difficult task. It can become an exercise in humility for the designer as he discovers these instruments' hidden virtues one by one, via time-consuming trial and error. Our intention in designing electronic displays is always, as it should be, to do the job better than it has been done before, but our first concern must be to assure that we do it as well.[6]

Following the Colgan 3407 accident at Buffalo, the ALPA noted the following in their submission to the NTSB:

> EFIS and the PFD is supposed to reduce the risks of visual illusions and spatial disorientation during instrument flight conditions by supplementing normal visual cues, but the tape system currently used is not intuitive. There is very little positional movement of the display due to the nature of the design, unlike a conventional analog airspeed indicator where experienced pilots can build appropriate mental models of airspeed based on position of the airspeed needle alone. The PFD airspeed tape appears static during the conventional instrument scan, and movement is not noticed unless focused attention is placed on the display, which is not a good information retrieval strategy when managing concurrent tasks during periods of high workload. Compounding this problem is the fact that the airspeed tape and altitude tape represent information in a radically different manner. If the airspeed tape is rising, airspeed is decreasing and a correction could require pitch down. Conversely, if the altitude tape is rising, the aircraft is descending and a correction would require pitch up. While this may be satisfactory to a pilot placed in an automation monitoring role, it is not intuitive to a pilot with who has predominant experience with conventional instrument displays and thus has that "embedded" in his or her mental model of the instrument scan.
>
> It is actually not appropriate to any pilot who is attempting to extract information and use it immediately, in a very, very high gain environment, to make complicated and timed control inputs.[7]

The Safety Board rebuked the flight crew of the Colgan airplane for failing to monitor the airspeed situation. They said, as Finding No. 8 in the final report, that

> Explicit cues associated with the impending stick shaker onset, including the decreasing margin between indicated airspeed and the low-speed cue, the airspeed trend vector pointing downward into the low-speed cue, the changing color of the numbers on the airplane's indicated airspeed display, and the airplane's excessive nose-up pitch attitude, were presented on the flight instruments with adequate time for the pilots to initiate corrective action, but neither pilot responded to the presence of these cues.[8]

A healthy effort to detect small, emerging failures, anchored with a rich awareness of discriminatory detail, would have tipped the crew off that something was amiss. The low-speed cue shouldn't have appeared when it did, if the ref speed was correct. It wasn't, of course, but that's the failure "elsewhere in the system" that should have been apparent. How does the low speed cue work? Why would it appear ahead of schedule? If you understand what its purpose is and how it works, that is, what its inputs are, then the indications are immediately interpretable.

Recall the reference we made to the Flight Safety Foundation's report in Chapter 7, regarding an accurate mental model and the impossibility of monitoring "a complex system if a pilot isn't sure how to correctly operate that system or what type of aircraft performance can be expected from each autoflight mode",[9] and consider that with respect to the arguments made by ALPA and by the Safety Board.

It is harder and harder to maintain a rich awareness of discriminatory detail when one is literally saturated in detail … and has emerged from airplane-specific training without an accurate mental model. Not only do we have a low-speed cue, we also have an airspeed trend vector as well as changing colors on the digits displaying indicated airspeed. All good information and all pointed toward something that should have disconfirmed or felt uncertain … but also a tremendous amount of detail to manage with a shrinking pool of Stanley Roscoe's residual attention. In fact, it may very well cloud the "richness" of the awareness. It would seem that Boeing's original focus in the design of electronic displays, "to do the job better than it has been done before, but our first concern must be to assure that we do it as well",[10] has disappeared from the industry's view, probably at around the same time we transitioned from the machine working in the service of man, to man working in service to the machine.

That said, it is important to understand that working hard at detecting small, emerging failures goes well beyond the contemporary application of active monitoring as a form of error trapping. The process of error trapping, really error correction, can and should, if it works right, take the emerging part right out of the equation, ejecting a small failure over the side. But this is a subset of failure; a significant portion of the whole set of failures does not involve error. Rather, it involves the detection of an unruly trajectory as described by Morin in his definition of the ecology of action, and that quite likely involves complexity.

Despite the automation developed over the past few decades, I still consider it wise to monitor a variety of system parameters. One particular day when operating a 737-800 from Denver to Miami, I did my regular top-of-climb survey of the electrical system. Lo and behold, I discovered the No. 3 TR was, supposedly, putting out not the usual 11 to 16 amps, but 161 amps. I knew this was, for all practical purposes, impossible. If it were true, we'd have smelled something burning by that time.

But something was up. I looked at the system diagrams and then looked at the generator outputs. At first, it looked like the right generator was putting out more amperage than usual … about 70 amps as opposed to 50. But later, the output stabilized back to a typical level. Since the right generator powers the right AC transfer bus, which powers the no. 3 TR, the power put out by the No. 3 TR has to come from the right generator. So there was no way this thing was cranking out that kind of amperage without a proportional increase in output from the right generator.

On the other hand, we're sitting at 37,000 feet with one hundred sixty passengers. Something is wrong, and it would clearly not be prudent to dismiss it the same way I might dismiss such an indication in, say, my car. The first officer and I decided it might be wise to run this by the technical folks, so I called the dispatcher and had him patch me through to the 737 tech rep on duty. The tech rep also could not accept that kind of power output, and he independently concluded that it must be an indication problem or a fault in the TR. In any case, his opinion was to continue close monitoring and write it up after landing.

That was what I expected to hear, but it was better to have actually heard it as opposed to assuming that I would have heard it had I not called. The next day, it turned out that the No. 3 TR had been replaced.

This is mindfulness, by Weick's definition, and prudence according to St. Thomas. But where does it leave us? We need to look closely to see if we can detect an emerging failure. We still have the possibility of one of Edgar Morin's unruly trajectories.

Contemporary safety programs are built around threat and error management. Within this structure are important aspects of human psychology … slips, lapses, violations, capture errors, illusions … all of which have played recurring roles in accident taxonomy. But the word "management" takes us right back to linearity, because management is predicated on some degree of predictability. We therefore need to consider the overlap between threat and error management and the ecology of action.

Consider a flight that I was scheduled to operate from Dallas to Albuquerque, which anyone in my old MD-80 fleet had done a hundred times. On that day, however, the ABQ forecast called for winds out of the southwest at between 20 and 30 knots and a broken cloud layer at 3,500 feet. Of course, there are no straight-in approaches to the west runways at ABQ; unless you are skilled in space shuttle operations, the mountains make straight in arrivals sort of problematic. There are a couple of RNAV approaches, however the MD-80 was precluded from flying RNAV instrument approaches at ABQ. Normally, this kind of wind would be handled by flying the RNAV Visual approach to runway 26. However, to do that you need visual contact with the airport at around 10,000 feet as you cross the Sandia Mountains; the broken ceiling wasn't going to allow that. The obvious solution was to fly an approach to runway 8, then circle to land on runway 21. Because the company did not have a training program for circling approaches, our Operations Specifications limited us to a ceiling of 1,000 feet and three miles visibility. The forecast met those minima easily.

However, the forecaster's written discussion for that day, contained in the National Weather Service Area Forecast Discussion, described in detail the major winter storm that was expected to churn across the New Mexico Mountains, which included travel advisories for all routes leading east and north of the city. In this case, the forecaster also said that there was some uncertainty as to whether the snow would impact the downtown area, although the current Terminal Area Forecast did not include any mention of that. The problem for us was FAR 121.613, which prohibits domestic departures unless the weather reports, forecasts, or any combination thereof indicate that the weather at the time of arrival will permit a landing. What I needed to keep an eye on was what kind of ceiling and visibility was actually going to develop during the day and how that might impact the circling maneuver. Two questions were open: would we be legal to depart on the basis of the circling approach and would we actually do it? Not the same question; I'm not inclined to circle an airplane weighing 130,000 pounds, at night, out of a strong tailwind, at 999 feet with only three miles of visibility, in snow, with rather tall mountains a couple of miles beyond the airport. But with a higher ceiling and better visibility, we could do it.

In the end, the city and airport were wide open, the wind at the surface died off, and we flew a standard ILS to a landing on runway 8. But throughout the day, while practicing mindfulness and prudence, we needed to also closely monitor the ecology of action. Each decision that we made represented an action that would immediately begin to escape our intentions. This would have been particularly true if, as we approached Albuquerque, it became clear that a circling approach would in fact be required. What "random incidents, chance, initiative, decision, the unexpected, the unforeseen, and awareness of deviations and transformations"[11] would be possible as we began that approach? Could such an approach enter into a sphere of interactions and be finally grasped by the environment in a way that may be contrary to the initial intention? You bet it could.

In the case of the screwy No. 3 TR, we were looking for the unruly trajectory of a systems interaction that no one had yet anticipated. We had obtained all of the counsel we could, judged of what we had discovered, and executed command. We remained ready to apply circumspection should the trajectory of the problem show any signs of further unruliness. In the case of the Albuquerque approach, we were doing the same thing, but also considering the overarching requirement to monitor the trajectory of our command decisions, looking for the slightest escape from our intentions. This brings us to the second aspect of Weick's concept of preoccupation with failure: working hard to identify mistakes that you don't want to make.

This process is ubiquitous in flight operations. We have a whole library of mistakes that we don't want to make. Things such as checklists, gear warning horns, stick shakers, and nearly every operational procedure that we use arise from the identification of significant mistakes that we do not want to make.

During World War II, my uncle was a fresh-faced twenty-year-old Army pilot serving as a C-46 instructor at Austin, Texas. He once told a wonderful tale about glider tow practice … at night. They were training both glider pilots and transport pilots in the art of towing troop gliders and this involved multiple tows, takeoffs, releases, and landings. After one landing, another glider was hooked up and my uncle and his student took off in the lumbering old 46. Becoming airborne, they discovered that the flaps were still in the landing position. As I recall, retracting the flaps on the C-46 was a complicated process, and apparently, they were not going to get it done immediately. He described the subsequent, fortunately successful, effort to stay flying while towing a glider with full flaps extended, and with a twinkle in his eye at the age of eighty-nine, looking over a cup of his favorite coffee, concluded with the statement, "that was the night we learned about checklists …".

His actions had clearly escaped from his intentions. The environment … the training environment, night operations, youthful inexperience, and a woeful absence of thoughtful risk management … had grasped the situation in a way that was most assuredly contrary to the initial intention. Sure enough, in today's operations, the process of verifying the correct takeoff flap setting, which can be very specific and often redundant, has developed after several well-known attempts to take off without the flaps or slats deployed.

From the good old days of flying the Atlantic with the inertial navigation systems in triple mix and even later with inertial reference platforms drifting slowly across the lower E&E compartment, the standard procedure has been to note your position over each fix, then capture the latitude when the longitude readout indicated you were two degrees past the last fix. Then you plot the latitude on that fix plus two-degree longitude … and the pencil dot is supposed to land on the line you have previously drawn between the last fix and the next one. Many pilots find this to be a boringly tedious task, but it is just a simple case of identifying a mistake we do not want to make. One can never be sure when one of those inertial systems is going to take a break, or, more likely, when the next fix that you typed into the box isn't the one you drew on the map. It doesn't take a lot of thought to realize that waiting until you get to the other side of the Atlantic to discover this kind of error can lead, at a minimum, to substantial egg on one's face. And, predictably, it has been done more than once.

Mistakes that we do not want to make are a subset of specific risk, but it is less a matter of looking harder for the lion crouching in the grass than it is avoiding the watering hole where the last five lion attacks have occurred. They can be used effectively to build boundaries and expand safety margins such that we have a better chance to spot and contain the trajectory of our small, but emerging, failure. Standard operating procedures are almost wholly designed around mistakes that we do not want to make; however, the capable and competent aviator can augment this body of knowledge with prudence that is specific to the flight and the situation.

Back in the Convair days, good old John Brown and I were holding outside of Bradley International Airport in Connecticut one morning. It was rather foggy that morning; visibilities were down below the Category I limit of RVR 1,800 feet. Another company airplane was also holding; at some point, the captain of that flight elected to depart the hold and proceed to his alternate based on fuel supply. Boston center issued a heading and off he went. We continued holding, as we were not at the bingo fuel point yet.

A couple of minutes later, Center called up our colleague and announced that the RVR had come up to minimums and asked whether he would like to return to Bradley for the approach. He did, and they issued him a new clearance back to Bradley. John looked over at me and said, quite emphatically in his Arkansas drawl (and I quote, because I have never forgotten his words), "That … was a mistake".

Our friend successfully landed at Bradley, as did we a few minutes later. I know we had plenty of gas left; I never did find out how much our colleague had when he landed. But if he had thought he needed to leave the holding pattern and divert, he must have been right around bingo fuel, or the quantity of fuel required to fly (a) from the hold to the IAF, (b) from the IAF to the MAP, (c) from the MAP to the alternate, and (d) forty-five-minute reserve. The mistake we don't want to make in this case is to use any of the fuel allotted for a, b, c, and d for the purpose of starting to divert and then returning to the original destination. If we do that, and then have to miss the approach and divert anyway, we end up arriving at the alternate with less than the minimum forty-five-minute reserve … possibly with no reserve at all.

And this was the point. Once our colleague had made the decision to abandon the divert and return to the original destination, his action entered into a sphere of interactions and would finally be grasped by the environment in a way that may be contrary to the initial intention, a sphere of interactions that may include the complex behavior of the cloud physics over the end of the runway. Without the margin of reserve fuel, he would not have had the resources to capture the unruly trajectory. This is a fundamental failure of strategy.

Through careful execution of the pilot-in-command's authority, flight-specific mistakes that we do not want to make can be identified and established as boundaries for the complexity that we may be encountering. The crew of Air Transat 236, which we discussed in Chapter 5, had a massive fuel leak suddenly occur over the Atlantic. Oddly enough, a rich awareness of discriminatory detail actually preoccupied them during the early period of the fuel leak. They had observed anomalous oil temperature and pressure indications on the right engine, and they became enmeshed in sorting this out, including discussions with company maintenance personnel via HF radio. It turns out that these indications were a direct result of fuel spraying across oil lines, but they had no way of knowing this. However, once a fuel imbalance advisory message appeared, a healthy preoccupation with failure would have immediately flagged a critical mistake that no one wants to make: screwing up a fuel leak over the North Atlantic. Procedurally, that mistake is clearly identified in the QRH text for fuel imbalance; however, the crew did not reference the QRH before cross-feeding fuel.

But the Air Transat crew really did not believe that a fuel leak was the problem, so it is possible that even if they had referenced the QRH procedure, they might not have followed the textual recommendation regarding a fuel leak. Rather than relying wholly on procedural guidance to identify mistakes that we do not want to make, the pilot-in-command must situationally anticipate and identify mistakes so critical that they establish a boundary of action. This boundary can be put in place early in a developing situation, and it can be used to protect the margins when faced with erratic, dynamic, and possibly complex inputs.

Arriving at the alternate with degraded reserve fuel is one such mistake; running out of gas over the North Atlantic is rather higher on the list. Through continuous awareness and careful monitoring, fuel discrepancies can be spotted early on. It really isn't enough to just check the fuel at each position report or at each waypoint. Rather, a comparison between fuel quantity indications and dispatch fuel minus total fuel burned should be made at those waypoints, and differential consumption, balance, and distribution should be checked frequently.

The identification of significant mistakes that we do not want to make can also be impeded when we are imbued with a mission-oriented attitude. We work on a daily basis with defined procedures that appear intended to preserve the mission, despite specific system failures, such as the procedures contained within the MEL and the QRH. But this appearance is a misinterpretation borne of an economic motivation. Both the MEL and the QRH are actually designed to manage the ecology of action. It is here that the normalization of deviance can enter the ecology of action and seriously degrade the strategy embodied in those documents, a strategy largely defined as avoiding mistakes that we have already identified as ones we do not want to make.

Finally, and in a similar vein, the identification of significant mistakes that we do not want to make can be critically degraded when we become enmeshed in Eddie Harmon-Jones' idea of action-based cognitive dissonance. Recall his definition, that dissonance between cognitions "evokes an aversive state because it has the potential to interfere with effective and un-conflicted action". Consider now what Weick is referring to when he describes "small, emerging failures", in other words, weak signals. How difficult is it to actively monitor, detect, and interpret weak signals when action-based dissonance absorbs whatever residual attention you have left? Looking back at the sequence of events leading up to the crash of FirstAir 6560 at Resolute Bay, the signals don't even appear to be particularly weak and yet they did not penetrate the captain's natural resolution of action-based cognitive dissonance.

We can get caught by a sense of mission orientation, by our natural drive to score highly on the "concern for production" axis of the Blake-Mouton grid. But by the time we encounter the action-based model of cognitive dissonance, time will be short and the acknowledgement of weak signals will require abandoning a clear plan and then formulating and executing a new plan with little to no time for contemplation. This can be mitigated by a strategy of default planning, for example, by vocalizing the initial steps of the published missed approach procedure as well as the standard go-around flow as we capture the final approach course. But it is vital to never underestimate how powerful action-based dissonance can be and how easily it can drown weak signals. By retaining a clear mental model of mistakes we do not want to make, we can create an opening for weak signals to be captured and acted upon.

And then there is the third aspect of Weick's definition of a preoccupation with failure: when people look for failures, they acknowledge the existence of incomplete knowledge.

Departing for Tel Aviv one night many years ago, flying a somewhat-fully-loaded Boeing 767-300 as TWA 884, we had both utility busses drop off line as we pushed the throttles up for takeoff. Being the savvy systems guys that we were, we assumed a spurious electrical surge of some sort, and I simply reset the busses. Takeoff competed, we retracted the gear … and the busses dropped off again. We reset them again. Shortly after, they dropped off once more. Now wising up, we got out the QRH, and found … predictably … that only one reset was allowed.

Climbing out toward Boston, we contacted our maintenance coordinator. During the next twenty minutes or so, we tried a variety of fixes. Finally, we found that the utility busses would stay on line if the left generator was taken off line. Maintenance saw no reason not to operate that way with the APU running. After all, one could operate with one engine generator deferred under the MEL.

We contemplated this as we headed toward Boston. I finally pointed out that we had no idea why the utility busses would not stay on line. However, when a generator is deferred properly, the CSD is disconnected and several other maintenance actions are accomplished. This was not the same situation at all. I told the captain and relief captain that, in my opinion, the NTSB Administrative Law Judge overseeing the hearing, at which we were trying to get our licenses back, would not have been impressed with our rationale, given that we had yet to pass Boston, let alone coast out over the Atlantic. They agreed; we turned back to New York. As it happened, a particular relay had fried itself. Not the best way to start a twelve-hour flight … never mind the multiple resets that were way out of bounds with the QRH procedure.

Consider how Weick's definition of a preoccupation with failure applied here. First, we had detected a small, emerging failure and it was indeed a clue to something else going wrong in the system. But then, allowing our mission-oriented drive to take over, we departed from the rigor of the QRH and entered into Morin's sphere of interactions, allowing our action to be fully exposed to the grasp of the environment. We really had no idea how our action, that of disconnecting the generator and re-connecting the utility busses, was going to interact with the actual problem that we had, the as-yet-unknown relay failure. Its behavior and interaction with other systems or components may have gone in several possible directions, leading to a result that was not even close to what we intended when we took the action. The clearly defined mistake that we did not want to make was that of entering an oceanic crossing with an unknown electrical problem, but a second, more insidious mistake that we did not want to make was to keep trying to reset a faulted electrical subsystem. But the third leg of Weick's definition comes into play as well. By applying the narrative of operating with a generator deferred, we were failing to acknowledge that our own knowledge of the system and the environment was incomplete. We really did not know what was causing the utility busses to trip off. Had we applied the full definition of a preoccupation with failure, we would have first detected the small, emerging failure. We then would have immediately grasped the mistake that we did not want to make, that of repetitively resetting an electrical subsystem that had faulted. Finally, we would have realized that we did not have complete knowledge of the system and its present environment. It took recognition of the later mistake that we did not want to make, that of starting across the ocean with a known electrical fault, before we captured what was clearly a wildly unruly trajectory. Had we opted to run the APU and continue, our action could very easily have been grasped by a sphere of interactions within the electrical system, leading to an outcome very much contrary to our original intentions.

But you may have noticed that we were practicing all kinds of good crew resource management. We were working together, and we even reached out to maintenance for their input. But here's the thing about that: as we learned so terribly painfully with the loss of Alaska 261,[12] the guys in the hangar will do everything they can to help … but all they can do is tell you what they would do in the hangar. They are assuming that you will not do anything that isn't an airworthy action. You, however, having slept through systems class, are assuming that they know the system better than you. That is yet another mistake that we don't want to make.

On January 9, 1998, United 965, a Boeing 767-300, was departing Zurich for Washington, D.C., when a problem developed with the center hydraulic system air-driven pump (ADP) during taxi out.[13] The aircraft returned to the gate and maintenance personnel eventually deferred the "AUTO" function of the ADP. Following takeoff, the crew received a "Leading Edge Slat Disagree" EICAS message. Communication with maintenance was initiated, and they suggested selecting the ADP to AUTO. The MEL deferral procedure actually required that the ADP be either selected to ON or OFF as needed, ergo, not to the AUTO position. However, selecting the AUTO position cleared the slat disagree message. Subsequently, the crew descended back to FL 200, slowed to flap speed, and extended the flaps to five degrees. It was confirmed that the flaps operated normally and that the leading edge disagree message was presented only when the ADP was in the ON or OFF positions, which were the two positions permitted by the MEL procedure. When selected to AUTO, the message disappeared. The crew elected to continue the flight with the ADP in the OFF position and the leading edge disagree message displayed.

Shortly thereafter, a second EICAS message appeared, indicating that the first officer's pitot heat was inoperative. The appropriate CB was pulled and reset (not a QRH procedure). It was noted at this time that two other breakers were tripped, one for the alternate flap drive motor and one for the passenger services outlets. The latter was reset but popped again immediately. Within a very short period of time, additional CBs began to trip and the first officer's EFIS displays flashed briefly, as did the engine displays. This pretty well cemented the idea in the crew's minds that they did not know what was going on. Nevertheless, they rather comfortably diverted to London Heathrow, rather than any closer possibilities, landed, and promptly filled up the first-class galley with smoke during the rollout.

It turned out that a galley chiller unit had been replaced incorrectly a day earlier, chafing some wires and leading to several instances of arc tracking. Recall that arc tracking is a complex physical system that is very much feedback-driven; rather notoriously, it does not necessarily pop CBs.

Once we have established robust monitoring, the next question is: what do you do with all of the data that comes flooding into your brain due to your rich awareness of discriminatory detail? You ask questions … repeatedly. Once you have become aware of detail that disconfirms, is unpleasant, feels uncertain, seems possible, is implicit, or is contested … you ask, why? Why is the torquemeter bouncing every time I feel a slight rumble in the back of my neck? Why is the dot I just plotted not on the line? Why on earth would the utility busses trip off consistently? Why would a configuration specifically prohibited by the applicable MEL item extinguish the asymmetry annunciation, but the two configurations that are required by the MEL will not extinguish the light? Why did I find two more CB's popped when all I did was pull and reset a different CB?

Both of these cases involved the dominance of a mission-oriented attitude, the incorrect use of crew resource management techniques, and willful violations of published procedures. In both events, a rigorous application of a preoccupation with failure might have led to better protection of the margins. Part of the reason for placing a high value on protecting the margins is that when an action escapes, becomes mistaken, the first couple of iterations are likely to remain within the margins. The margins are limited, however, and this, of course, is the whole point behind a preoccupation with failure and looking for an unruly trajectory.

However, the ecology of action entails a host of thoughts that we haven't thought of yet. This is really Morin's point about the sphere of interactions. While the boundaries that we construct through understanding mistakes that we do not want to make, and the associated margins, can contain many unruly trajectories, the existence of complexity, local interactions, and open systems with a high degree of sensitivity to initial conditions means that there are plenty of unruly trajectories that we still *will never, ever see coming.*

But both events also offer an opportunity to revisit St. Thomas and his dissertation on prudence. What kind of counsel was taken? In both cases, it was significant, particularly from the maintenance folks. But when we get to the second tenet of prudence, "to judge of what one has discovered" … there lies the weakness of a mission-oriented, "git 'er done" type of approach. When Thomas was talking about caution, he was quite accurately talking about risk management. As we discussed in Chapter 3, he even split up his definition of caution between "some are of frequent occurrence; the like can be grasped by reason, and against them caution is directed"[14] (specific risk) and "Others there are that occur rarely and by chance, and these, since they are infinite in number, cannot be grasped by reason …"[15] (probabilistic risk).

In both cases, whether it be continuing a flight with utility busses that won't stay on line unless a generator is taken off line or continuing a flight with an unexplained but erroneous leading edge slat warning, the risk was initially assessed as a specific risk, one that can be grasped by reason. In both cases, the actual risk was more probabilistic, insofar as there existed multiple, unknown probabilities of unknown failures. In both cockpits, there was no way to measure these probabilities; they clearly would occur rarely and by chance and could not, with the available information, be grasped by reason. The initial approach of both crews was to box these events into a template of specific risk by applying a known narrative, which is the opposite of a reluctance to simplify and negates a preoccupation with failure. In at least the United case, it is likely that a certain sprinkling of the normalization of deviance was involved, in terms of flight crews building a great deal of experience with faulty indications.

At the end of the day, we in the cockpit are the ones who must judge. Everyone else who is offering counsel is doing so on the presumption that we will, in fact, judge of what we have discovered. There is no one else to do it.

NOTES

1. Karl Weick, Kathleen Sutcliffe, *Managing the Unexpected: Assuring High Performance in an Age of Complexity*, Wiley. Kindle Edition, 2015, 46.
2. Karl Weick, "Reflections on Enacted Sensemaking in the Bhopal Disaster" *Journal of Management Studies*, Vol. 47, No. 3, May 2010, doi: 10.1111/j.1467-6486.2010.00900.x.
3. Karl Weick, Kathleen Sutcliffe, "Managing the Unexpected", 2nd Managing the Unexpected in Wildland Prescribed Fire Workshop, Jacksonville, Florida, February 28, 2005.
4. Ibid.
5. David K. Graham, "Electronic Display of Powerplant Parameters", *Society of Automotive Engineers*, No. 841467, 1985.
6. Ibid.
7. Air Line Pilots Association, Submission to the National Transportation Safety Board Regarding the Accident Involving Colgan 3407, DHC-8-Q400, Clarence Center, New York, February 12, 2009.
8. National Transportation Safety Board, AAR-10/01, Loss of Control on Approach, Colgan Air, Inc. Operating as Continental Connection Flight 3407, Bombardier DHC-8-400, N200WQ, Clarence Center, New York, February 12, 2009, 151.
9. Flight Safety Foundation, "A Practical Guide for Improving Flight Path Monitoring, Final Report of the Active Pilot Monitoring Working Group", November, 2014.
10. Graham.

11. Edgar Morin, *Seven Complex Lessons in Education for the Future*, United Nations Cultural Organization, November 1999, 45.
12. National Transportation Safety Board, AAR-02/01, Loss of Control and Impact with Pacific Ocean, Alaska Airlines Flight 261, McDonnell Douglas MD-83, N963AS, About 2.7 Miles North of Anacapa Island, California, January 31, 2000, 4.
13. United Kingdom Air Accidents Investigation Branch, Aircraft Accident Report 5/2000 (EW/C98/1/3), report on the accident to Boeing 767-322ER, N653UA at London Heathrow Airport on January 9, 1998.
14. Thomas Aquinas, *Summa Theologica*, translated by Fathers of the English Dominican Province, e-artnow Kindle Edition, 2013, IIa-IIae,, q.49, a.8, ad.obj.3.
15. Ibid.

9 Reluctance to Simplify

As a ground instructor, I always thought the goal was to simplify aircraft system functions, in concept, in order to explain what the thing was doing. I did, and still do, prefer the use of a white board and marker to sketch out system schematics in simplified, cartoonish form, as opposed to projecting a blown-up schematic from the aircraft maintenance manual. I periodically encounter former students from years gone past, and for some reason, my ideas about simplification do not seem to be a part of their memory of our classroom time together. Nonetheless, a good instructor needs to know how and when to simplify concepts in order to build understanding.

Linear systems are usually pretty easy to simplify because they are so reliably predictable and reducible. Even chaotic systems can be simplified through the use of strange attractors, which can model the system's behavior in general, if not precise, terms over time. Complex systems, which are irreducible to their constituent components, are pretty resistant to simplification.

Nevertheless, simplification remains a powerful tool in the development of understanding. We are used to it. We use it every day in the process of sensemaking, which is nothing more than how we interpret the world through the narrative mode of thought. We project the outcome of the unfolding narrative, as yet incomplete, by using experience and technical comprehension to select one of the known narratives in our memory, or perhaps just generating an abductive hypothesis, and applying it to the unfolding narrative in an effort to predict what happens next.

So the notion of a reluctance to simplify is a challenging and contradictory idea to get our arms around. Not only are we living in a world that increasingly gravitates toward a black and white view of almost any form of thought, but on the face of it, a reluctance to simplify can seem quite the opposite of standardization.

Weick points out that "simplification obscures unwanted, unanticipated, unexplainable details and in doing so, increases the likelihood of unreliable performance".[1] This was precisely what we were doing while tootling toward Boston with a pair of recalcitrant utility busses. The simplification was that by taking a generator offline in order to get the utility busses back on line, we were replicating an operation with a generator deferred under the MEL. A reluctance to simplify would acknowledge that we were doing no such thing; the unexplained detail was that we had no idea why the utility busses would not stay on line with the left generator, and the unwanted detail was that the required maintenance procedure for deferring a generator had not been accomplished.

On the other hand, it is easy to draw the conclusion that investing attention in unwanted, unanticipated, or unexplainable details elevates the chance of error, and reducing or eliminating such error is the goal of standardization. In this paradigm, the correlation between standardization and simplification seems obvious, but only

DOI: 10.1201/9781003369677-9

if you equate the words unwanted, unanticipated, or unexplainable with the word *unimportant*. In truth, standardization is not, by default, simplification, even if the result is a movement toward the constraint of variation.

The problem is how we tackle thinking about those unwanted, unanticipated, or unexplainable details. We are stuck with the narrative mode of thought here; the details must fit into a story, and with only a partial listing of those details, we must choose from a selection of narratives defined by our previous experience in order to make sense of the situation. It is this sense-making that employs simplification; been there, done that, got the tee shirt … and it still fits. Sense-making is a function of experience modified by technical comprehension, and it can be heavily influenced by the normalization of deviance. Our experience with utility busses dropping off when we pushed the power up was influenced by deviance normalized through the common behavior of older airplanes. For example, the MD-80 has a known propensity for spontaneously popping circuit breakers if the external power is pulled without changing over to the APU first, which can tend to normalize strange electrical system behavior in your mind when power surges or spikes are perceived or imagined. On top of that, the 767 utility busses *will* drop off when you push the throttles up … if the engines are shut down. That said, I have never seen the utility busses drop off on a 767 when the power is pushed up for takeoff, either before that night or since.

Yet in the pursuit of error reduction when operating a large, computerized airplane in a three-dimensional environment completely enveloped by an atmosphere that is, from the ground up, the definition of a complex system, we have leaned heavily on simplification as the default interpretation of standardization. This gravitates toward an imbalance. Standardization is a tool with which to remain within the initial space of safe operation, not a tool to make life simpler. Interpreting it as simplification only drives your sense-making beyond the limits of its value, because you become overconfident in your selection of the narrative that you believe will complete the situation.

Nonetheless, simplification abounds. This can be traced back as far as Cessna's post-World War II ideas about simplifying flight training so that anyone could learn to fly (why do we need to do spins, anyway?), through designs such as the Ercoupe which removed pilot inputs to the rudder, emerging today in the deep, deep layers of automation intended to markedly simplify the choices and inputs available to the pilot (while introducing substantial complexity to the system as a whole). One of my favorite cartoons shows two cockpits in comparison: in one, a DC-9 pilot is sitting in a cockpit that looks more like the engine room of a World War I U-Boat, and in the other, an A320 pilot is sitting in a cockpit that has a single lever marked "Up/Down".

The drive toward simplification can be mesmerizing; it distracts us from not only the complex environment but also from the complexity introduced by Morin through the ecology of action. This is easily spotted in such bumper-sticker ideas as "Every Day is Groundhog Day". Simplification implies that the system will keep the situation from exceeding the bounds of your comprehension and awareness, when in fact it will do no such thing. You can get a start on becoming reluctant to simplify if you begin to realize that only one day a year is Groundhog Day, and you're actually not Bill Murray stuck in a movie.

The imbalance generated by interpreting simplification and standardization interchangeably, and the effects of complexity, is probably best described by the philosopher Gilbert Ryle, when he said that

> To be thinking about what he is here and now up against, he must both be trying to adjust himself to just this present once-only situation *and* in doing this to be applying lessons already learned … it is the pitting of an acquired competence or skill against unprogrammed opportunity, obstacle or hazard.[2]

The balance between applying acquired competence against un-programmed opportunity, obstacle, or hazard is the balance between sense-making and sense-discrediting. The term "sense-discrediting" refers again to the application of a rich awareness of discriminatory detail … mindfulness … to the present once-only situation while adhering to the template of standardization. It requires what Eric-Hans Kramer has called "real openness",[3] which goes beyond simply being open to new information, to the idea of being open to information that *we have never thought of before*. In other words, we need to be open to the possibility of a black swan.

Each approach to landing is, in many, many ways, the same. Each approach to landing is, also in many ways, different. To apply a reluctance to simplify during an approach is to fly the stabilized approach profile while maintaining a rich awareness of discriminatory detail … differences between, say, tower-reported wind and FMC wind, between runway visibility and flight visibility, between the power the autothrottle should be using and that which it is using, or between the barometric altitude while descending the glide slope and the radar altitude. Sense-discrediting circles back to the basic tenet of mindfulness: we are looking for things that disconfirm. We are essentially trying to falsify our sense-making.

At 1,500 feet AGL, the Boeing 767 autoland status annunciator switches to "LAND 3". This is an indication that the autoflight system has satisfied all required criteria for an autolanding with three fully operative autopilots. Strangely enough, obtaining a complete list of what the autoflight system has actually looked at before displaying a "LAND 3" annunciation is remarkably hard to come by … the people who write approved manuals seem to have no reluctance to simplify whatsoever. And, indeed, many training scenarios emphasize little more than monitoring the "LAND 3" annunciation and being prepared to go around if it drops back to "LAND 2" or to "NO AUTOLAND". How might one approach this through a reluctance to simplify?

Well, to begin with, the autoland status annunciator doesn't say anything about the status of ground facilities. In July 2000, Air New Zealand 60, a Boeing 767, began a coupled approach to Apia, New Guinea.[4] All three autopilots were engaged with LAND 3 displayed. However, due to a complicated chain of errors involving maintenance of the ILS ground equipment, the actual glide slope was not operating; only a carrier wave signal existed. The airplane was quite happy to descend with this signal as guidance, and rather steeply, as it turned out. Using a rich awareness of discriminatory detail, the first officer noted the unusually close proximity of island lights, which seemed oddly out of place. The captain simultaneously recognized a DME distance vs. altitude discrepancy, and they executed a go-around from

400 feet AGL while still 5.5 miles from the runway. Thankfully, the flight crew's awareness had been elevated during their crew briefing by a NOTAM stating that the ILS equipment was unmonitored; this probably enhanced their enthusiasm for sense-discrediting.

An awareness of NOTAMs, and understanding that LAND 3 refers to aircraft systems only, and a sequential crosscheck of altitude against DME are techniques that resist simplification. In particular, crosschecking the altitude against the DME is an example of sense-discrediting, a method for questioning what appears to make sense, that is, LAND 3 with glide slope and localizer centered.

A couple of years ago, we were descending into Chicago for a landing at O'Hare. In fairly typical Chicago fashion, we were assigned two different arrivals and three different runways within twenty miles of the airport. Our final runway assignment was to 10R. We tuned the localizer, identified it, intercepted the final approach course, configured, and tootled down the glide slope. Not a cloud in the sky, all of Chicago lit up before me … but no runway. I'm coming down through about 2,000 feet and I don't see a runway. I see the north complex, and the terminal … but no runway 10R. Needles all centered up, and I'm thinking, what is going on here? I'm approaching the world's most cantankerous airport and have no idea where the runway is. Suddenly, it dawned on me, and glancing quickly over to the approach plate, I spotted what I thought I might … a little tiny note about an offset localizer. Sure enough, and there's the runway right over there. That's what happens when you simplify a visual approach briefing … although exercising a reluctance to simplify in the rather dynamic world of Chicago airspace can be challenging.

A very common area of rather spectacular simplification is the weather. As we considered in Chapter 5, weather phenomena are largely higher-level features emergent from a complex system of air and water molecules self-organizing one way or another. But as pilots, we are generally handed a TAF or two along with a couple of color weather charts that have devolved almost to the level of a cartoon; they contain little specific information and zero nuance. There is, of course, a great deal more weather information available, but it takes time to study.

While sitting in the cockpit at St. Louis, my first officer and I were looking at the TAF and METAR history for Dallas one afternoon. There was a lazy stationary front draped across South Texas. It had been there for days; to the north was a very, very steep inversion sitting at around 2,000 feet, extending well up into Kansas and Nebraska. The inversion had been the instigator of quite a band of freezing rain and drizzle over the past few days. In the Dallas area, the ground was soaked and the temperatures above freezing, but the inversion kept the humidity trapped. The place was saturated in fog, about which we didn't really care except that our airplane that day had been downgraded to Category I approaches only. The TAF earlier in the day had forecast an improvement from the 1/8 mile, 1,200-foot RVR that was parked at Dallas. But it wasn't happening. The TAF kept getting updated, pushing the improvement time back to later in the afternoon.

As with my Albuquerque example in Chapter 8, the problem was again FAR 121.613, which prohibits domestic departures unless the weather reports, forecasts, or any combination thereof indicate that the weather at the time of arrival will permit a landing. Since we were restricted to Category I, we really needed to see some reliable evidence of the improvement. The lag in improvement called into question how much trust we should place in the forecast. As our departure time approached, the RVR at Dallas was still 1,400 feet.

It would be easy to simply take the updated TAF at face value, but that isn't how the FAA sees it. Their interpretation of 121.613 clearly says that while a combination of reports and forecasts can be used to conclude that improvement will occur by the estimated arrival time, it also means that a combination of reports and forecasts can lead to a conclusion that conditions will *not* improve even if the current forecast says they will. What we needed to do was look into the weather situation a bit more deeply and see what "unwanted, unanticipated, or unexplainable details" might exist, and which way they were pointed.

The forecaster's discussion that morning had also included an extensive discussion of an approaching low pressure system and the energy it would bring into West Texas that day. This discussion provided details of wind shifts and dew point changes that would precede the front as it started to move north, along with an expectation of two bands of convective activity that would develop ahead of the incoming low. Both of these characteristics were evident in the current METARs and radar plots. This gave us confidence that the current TAF, which really looked the same as all of the preceding TAFs that day, would indeed prevail and the below-minimum RVR's would not persist to our estimated time of arrival. And indeed, they did not.

But we must stop and consider a critical point regarding our capable and competent aviator, which is the remarkably prescient description of the extroverted pilot written by the CAA back in 1939. Recall that they said that such a pilot candidate possessed "such features in their daily lives as simultaneous interests in a number of business enterprises, frequent entertainment, lack of interest in abstract thought, hearty dispositions and carefree attitudes".[5] This set of characteristics do not easily lend themselves to such things as a reluctance to simplify. The carefree attitude doesn't help, and the lack of interest in abstract thought is nearly fatal to an understanding of complex systems.

Complex systems are abstract in the sense that they cannot be understood by disassembling them into their component structures and examining those components with a linear, Newtonian template. Recall that complexity is a behavior that is rich, collective, and dynamic, created from simple interactions between very large numbers of components. Recall also that a complex system is extremely sensitive to initial conditions. As such, they are inherently unpredictable, and this is why mindfulness, falsification, and sense-discrediting are so important.

Sadly, one area where simplification has had an enormous impact is in the area of flight training. Ever since Cessna came up with the idea that anybody could learn to fly, which of course would sell more airplanes, the industry has been on a mission to simplify training. Some years ago, the mathematician, Peter Hilton, drew a useful distinction between training and education, and pointed out that many people conflate the two by perceiving a mathematics education as the acquisition of a defined set of skills, applicable to future endeavors. This acquisition is largely associated with memorization, as opposed to genuine understanding. Yet it is the ability to adapt to change that is the "hallmark of a successful education". A training, according to Hilton, contributes little to education and yields only brief advantages.[6]

The parallel between Hilton's thinking and Ryle's "once-only situation" and the application of "lessons learned" is clear. Pilots should be educated as well as trained. Training alone induces simplification; to practice a reluctance to simplify, an education must be present as well.

Let's consider how both a reluctance to simplify, a preoccupation with failure and a genuine mechanical education might have influenced the Air Transat engine change crew in Montreal. Bear in mind that one of the most pernicious forces driving simplification is schedule pressure. This crew of mechanics had an operational deadline that was fast approaching. This is a form of top-down feedback into a complex system, reaching all the way into the highly complex functions within each mechanic's brain. It is not an easy factor to mitigate.

That said, a reluctance to simplify would have elevated the consideration of actually getting their hands on the applicable service bulletins, in order to ensure that the procedure was not haphazardly simplified. While it was certainly simpler to only replace the three fuel tubes, a preoccupation with failure, upon encountering the interference between the new fuel tubes and the existing hydraulic lines, should have initiated the reluctance to simplify. Such a reluctance could have forced a proper consideration of the lack of access to the correct service bulletins. The technician reconnecting the original, pre-modification hydraulic tube noted that it had a tendency to "spring back", interfering with the required clearance with the newly installed fuel tube. This was resolved by positioning the hydraulic tube and then tightening its "B" nut so that it retained the correct clearance. The "spring back" behavior was a failure; preoccupation with such failures would have triggered an alarm in everyone's head that something was amiss. Does anyone think that an acceptable design would tolerate such a condition? Finally, a deeper education in mechanical systems and their structures should have caused the technicians to realize that, while the "B" nut successfully held the tube in position, this left the tube pre-stressed mechanically, and the additional force of applied hydraulic pressure would easily combine with that pre-stressing to overcome the "B" nut. Instead, a superficial level training had been applied earlier in their careers and that training made no specific mention of such a possibility.

In the Air Transat case, simplification very much obscured unwanted, unanticipated, and unexplainable details. Likewise, the crew I was a part of while hurtling toward Boston with the recalcitrant utility busses was experiencing a high-level, top-down feedback pressure, as we were already airborne and headed for a long night of flying. The maintenance control folks were offering a simplification of the problem by suggesting that it was no different than a generator deferred by the MEL; indeed, I subsequently had a rather heated discussion with another captain on this very incident over that exact simplification. Only through a preoccupation with the particular failure, and a consideration of what we did not know, were we able to deploy a reluctance to simplify this situation.

NOTES

1. Karl Weick, Kathleen Sutcliffe, *Managing the Unexpected: Assuring High Performance in an Age of Complexity*, Wiley. Kindle Edition, 2015, 64.

2. Gilbert Ryle, *On Thinking*, Konstantin Kolenda, Editor, Rowman & Littlefield, 1980, 129.
3. Weick, 68.
4. Civil Aviation Authority of New Zealand, Aircraft Serious Incident Report, Occurrence Number 00/2518, B767-319ER, ZK-NCJ, Faleolo Airport, Samoa, July 29, 2000.
5. John W. Thompson, "Studies to Develop Methods of Determining Emotional Stability", Civil Aeronautics Authority, Bureau of Federal Airways Technical Development Division, Note No. 18, Washington, D.C., 1939.
6. Jan Gullberg, *Mathematics, From the Birth of Numbers,* W. W. Norton, 1997, xvii.

10 Sensitivity to Operations

Karl Weick defined "operations" as the "site where the realities of unlimited interdependencies and limited concepts meet in real time".[1] In other words, where the real world meets management's expectations. We used to have a shorter definition ... it's "where the rubber hits the road".

Unlimited interdependencies refer to Morin's ecology of action. Limited concepts take us back to St. Thomas's observation that "the need for reason is from a defect in the intellect",[2] insofar as we are not angels and thus do not comprehend everything by simple insight. If we consider the ecology of action meeting the defect in the intellect, we've got a pretty good map of the operations topography.

On December 20, 1995, American 965, a Boeing 757, hit the top of a mountain near Cali, Columbia.[3] Approaching from the north, the crew had planned a landing on runway 01; however, they accepted the controller's offer of the approach to runway 19. This created a rushed environment. They were high and had not briefed the approach to runway 19 nor loaded it into the FMC. Because of considerable confusion in the communications with the controller, confusion about their present position, and an egregious duplication of navigational fix identifier codes, they ended up flying the airplane into a mountain top.

I need to pause for a moment and reflect a bit on this particular crew. They had the spoilers deployed for the rapid descent, and when the GPWS started screaming, they responded absolutely appropriately; max power, pitch up to the stick shaker, etc. But they never stowed the spoilers. This was perhaps the most immediately significant mistake they made; had the stowed the boards, they most probably would have cleared the terrain.

About ten years after this accident, I was getting a six-month checkride in the simulator as a current 757 captain. We had a very creative and skilled check airman running this. He had us out in the Midwest somewhere, I don't recall just where, but it was flat as a pancake and we knew it. He had us in some kind of rapid descent with the spoilers up; we were still at around 15,000 feet. There we were, coming down like a safe, and out of nowhere the check airman triggered a GPWS terrain alert.

"CAUTION! TERRAIN! TERRAIN, TERRAIN! PULL UP! PULL UP!"

I had no idea what was going on or what possible terrain we could be approaching. My first thought was, what is he doing? There's nothing out here. It was absurd. Nevertheless, being a good student, I immediately disconnected the autothrottle, firewalled the thrust levers, and pitched up to around twenty degrees.

And left the spoilers deployed.

In popular culture, we often hear about simulator sessions so realistic that the pilots emerge all soaked in sweat and trembling; most of us know that is utter nonsense. It's a bloody simulator, everybody knows that, and no one ever confuses the sim for the real airplane the way Martin Sheen confused a ceiling fan for a helicopter. And yet, while I had never once climbed out of the sim feeling anything

DOI: 10.1201/9781003369677-10

close to emotionally traumatized … I did this time. He got me and he got me good. It was one of the most valuable simulator experiences I have ever had. I knew instantly that I had just made the identical mistake as the crew of American 965. That was unbelievably humbling.

So let's analyze what happened, but let's always … with this accident and any other … remember that any of us, all of us, can make similar mistakes.

The Cali approach controller did not have radar. He needed to rely on position reports from the aircraft. He also did not speak English beyond the minimum necessary for air traffic control communications. The original clearance had been to the Tulua VOR (ULQ). However, the controller then *seemed* to clear the flight to fly direct to the Cali VOR. Actually, he didn't; he cleared them to the Cali VOR along the present routing, which would still include the Tulua VOR. However, the captain interpreted the clearance to be direct to Cali, read it back that way, and received no correction from the controller. When the captain loaded that routing into the FMC, the Tulua VOR was deleted. At this point, the controller offered the approach to runway 19. After some discussion with the controller, the crew of 965 requested routing direct to the Rozo NDB for the transition to the approach and controller simply repeated his clearance to fly the Rozo One arrival. However, he still insisted on a position report over the Tulua VOR. At this point, the crew began to experience significant confusion; they were already past Tulua, but did not realize that. Consequently, when the captain tried to reload Tulua into the FMC, it would not appear on the navigation display. Their confusion was clearly identified in the report based on the CVR transcript:

> Their inability to effectively do both tasks is evidenced at 2138:49, when the first officer asked, "where are we", followed by a short discussion between both the captain and first officer regarding their position relative to the ULQ VOR. Again at 2139:30, two minutes before impact, neither flightcrew member could determine which navaid they were to proceed towards. The first officer stated, "left turn, so you want a left turn back around to ULQ". The captain replied, "Nawww … hell no, let's press on to …" The first officer stated, "well we're, press on to where though?" The captain replied, "Tulua". The first officer said, "that's a right u u". The captain stated, "where we goin'? one two. come to the right. Let's go to Cali first of all, lets, we got [expletive] up here didn't we". The first officer replied, "yeah".[4]

At this point, we should recall St. Thomas definition of circumspection, one of the critical elements of prudence:

> Just as it belongs to foresight to look on that which is by its nature suitable to an end, so it belongs to circumspection to consider whether it be suitable to the end in view of the circumstances.[5]

This is the core principal of a sensitivity to operations. Weick elaborates on this idea by citing Gene Rochlin's 1997 description of the term "having the bubble":

> Those who man the combat operations centers of US navy ships use the term "having the bubble" to indicate that they have been able to *construct* and *maintain* the cognitive

> *map* that allows them to integrate such diverse inputs as combat status, information sensors and remote observation, and the real-time status and performance of the various weapons and systems into a single picture of the ship's overall situation and operational status.[6]

This can be interpreted as situational awareness. However, like Weick, I believe that the term "having the bubble" is a more specific term. It really encompasses both situational awarenesses: the old one, where are we, where is the ground, where are we going, and at what angle of attack, in parallel with the new one, where does the autoflight system think we are, where does it think we want to go, how is it going to get us there, and where are we within one or more flight envelopes that it is supposed to keep us from departing? Clearly, the crew of American 965 had not been able to construct a cognitive map, much less maintain one, and they weren't doing very well at integrating diverse inputs.

Consider mindfulness. The accident sequence seems to be an explicit case of a crew concentrating on those things that "confirm your hunches, are pleasant, feel certain, seem factual, are explicit, and that others agree on".[7] And yet, from the CVR transcript cited above, we can spot plenty of things to concentrate on that "disconfirm, are unpleasant, feel uncertain, seem possible, are implicit, and are contested".[8]

What about a preoccupation with failure? From the Columbian report: "At 2140:40, the captain stated, 'it's that [expletive] Tulua I'm not getting for some reason. See I can't get. OK now, no. Tulua's [expletive] up'". This is clearly a failure; it is important to consider that the crew felt rushed, and under such circumstances, a preoccupation with anything can be very distracting and quickly dismissed. However, a mindset constructed around a preoccupation with failure would, perhaps, have hoisted a red flag immediately when the problem with Tulua emerged. It should have been in front of them still, or so they thought, but it wasn't. How did that failure occur?

Quickly dismissed, indeed; consider the role of cognitive dissonance. The Columbian report put it this way, citing Wickens (1984):

> Operators tend to seek (and therefore find) information that confirms the chosen hypothesis and to avoid information or tests whose outcome … could disconfirm it. This bias produces a sort of "cognitive tunnel vision" in which operators fail to encode or process information that is contradictory to or inconsistent with the initially formulated hypothesis. Such tunneling seems to be enhanced particularly under conditions of high stress and workload.[9]

This is really just a re-stating of Eddie Harmon-Jones' concept of action-based cognitive dissonance, in which he suggested that dissonance between cognitions "evokes an aversive state because it has the potential to interfere with effective and un-conflicted action".[10] In this model, the crew of American 965 was very much in an action-oriented frame of mind, and thus dissonant feelings were almost certainly going to be dismissed in order to resolve the inherent stress resulting from the dissonance. In this "action-oriented" state, immediately following the decision to fly the approach to runway 19, the reduction of cognitive dissonance will be at a premium in order to take "effective and un-conflicted" action and implement the decision.

A sensitivity to operations would have flagged several aspects of the situation. First, the change in approach was unexpected. Second, they were going to have to rush to get down. Third, there is no radar and the controller's English cannot be relied on. Finally, the crew is probably fatigued. Under normal circumstances, the failure to get Tulua loaded might not demand a high degree of preoccupation. However, in the context of a sensitivity to operations, that failure becomes vastly more significant. It should have raised a red flag.

On August 14, 2013, UPS 1354, an Airbus A300, crashed on approach to Birmingham, Alabama.[11] This accident was not that unlike the Cali crash; once again, the crew was faced with a sudden change in runways. The longer runway at BHM, runway 6, was closed and would reopen just a few minutes after they arrived in the terminal area. The controller offered the non-precision localizer approach to runway 18, which they accepted.

Unfortunately, while they correctly captured the localizer, they failed to re-sequence the waypoints in the FMC. Because of that, the FMC was still headed straight to the center of the Birmingham Airport and it failed to capture the vertical profile as they crossed the final approach fix. This is an example of the two situational awarenesses; the airplane was going in one direction while the FMC thought they were going in another direction.

Recognizing that something was amiss, the captain switched to a constant descent angle procedure for the approach, utilizing vertical speed while tracking the localizer inbound. Initially, he selected 700 FPM down; shortly after that, he increased the vertical speed to 1,000 FPM and ultimately selected 1,500 FPM. The first officer thought that the FMC had failed to capture the profile because they were too high. In fact, they had crossed the final approach fix only two hundred feet high.

But in reality, neither of them had any idea where they were on the vertical profile. They were still in the clouds. When the EGPWS started yelling "SINK RATE" at them, the captain dialed back the vertical speed to 600 FPM and spotted the runway just as he did. He reduced the vertical speed further to 400 FPM … and promptly flew into the trees.

When they crossed the final approach fix, two hundred feet high already, and the autopilot failed to capture the vertical profile, they ran headlong into the unruly trajectory resulting from the ecology of action. The trajectory was already off on a tangent because they had failed to re-sequence the waypoints. Now, the captain attempted to chase the unruly trajectory and bring it back under control by resorting to a constant descent angle technique. But that isn't really what he did. Instead, he essentially adopted a dive-and-drive technique, perhaps without really realizing it, more or less racing down the vertical profile in order to reach the MDA before the missed approach point. But the ecology of action was in full swing; the descent technique had not been planned or briefed, they had started the descent slightly high and slightly late, nobody really knew where they were on the vertical path … and nobody called out the MDA. Pretty much every action this crew took escaped their intentions.

A sensitivity to operations really ties together everything we have discussed so far. The front door to a sensitivity to operations lies with the NOTAMs. The NTSB criticized the dispatcher for failing to notify the crew of the runway closure. Yet it was a NOTAM in their flight release paperwork. Had they accessed that information before they left Louisville … part of the first act of prudence, taking counsel … they would have had plenty of time to formulate a change to the strategy.

During the approach briefing, the captain pointed out the non-standard glide path angle of 3.28 degrees. But he did not elaborate on the requisite descent rate. A sensitivity to operations would suggest discussing this unusual condition; the required descent rate, at their approach speed of 140 knots, would have been a bit over 800 FPM. In itself, this isn't particularly dramatic. However, it is 75 FPM greater than that required for a standard three-degree glide path and close enough to the mandated maximum of 1,000 FPM incorporated in the stabilized approach criteria to warrant a briefing on under what conditions, for how long, and how much beyond 1,000 FPM might be necessary. The recognition of this approach as different from a standard approach might have highlighted the threat posed when the autopilot failed to capture the profile.

Of course, they were expecting that the autopilot *would* capture the profile. Here we see the seeds for the action-based case of cognitive dissonance. The captain … really both of them … were in an action-based frame of mind. When the profile did not capture, a preoccupation with failure could have triggered a sense of things which "disconfirm, are unpleasant, feel uncertain, seem possible, are implicit, and are contested".[12] There are, in fact, several mistakes here that we do not want to make. First, using an approach technique … either a constant angle descent or a dive-and-drive … that has not been briefed and coordinated. Second, most importantly, beginning an already steep descent profile later … and higher … than the final approach fix. Third, allowing the approach to de-stabilize in order to catch up to the profile, a profile that really is no longer defined by any airborne guidance. But the action-based version of cognitive dissonance resolves that dissonance by suppressing or extinguishing such concerns.

It must be borne in mind that the last weather observation the UPS crew received indicated a broken ceiling of 1,000 feet. The captain was almost certainly thinking that he needed to get down quickly so as to be at least close to the vertical profile when he broke out, at which point he would have plenty of time to ensure a stabilized approach.

I have done something similar. Approaching Monterrey, Mexico, one night, we were aware of a scattered layer at four hundred feet, a broken layer at eight hundred, and rain. The approach was offset. The MDA was about four hundred feet above the ground. I had briefed the first officer that, due to the scattered layer, I suspected we might not spot the runway until it was too late to safely maneuver into a position to land. Our MD-80 had no VNAV capability, so this was very much a constant descent angle approach, and it is also a slightly steeper approach path, at 3.34 degrees. And as it happened, I was slightly late in initiating the descent. To compensate, I rolled in a 900 FPM descent rate. I was thinking that I wanted to get down and break out sooner rather than later, because I really wanted time to evaluate the visual situation once we picked up the lights. I can easily imagine the UPS captain thinking the same thing.

When we got to the MDA, we saw … nothing. Not a single light, or feature, or anything. It looked exactly the same as it had at 5,000 feet. Our procedure precluded us from setting the missed approach altitude until we sighted the runway, and so the autoflight system would capture the MDA if we had not yet reset the altitude window. If it captured, we were obligated to miss the approach, which we promptly did. I was flummoxed; I had expected to see something, but perhaps not enough to land. I did not expect to see nothing whatsoever.

Here is a second point where a preoccupation with failure could have made a difference. Thirteen seconds before impact, and after they had passed the MDA, the UPS first officer said, "it wouldn't happen to be actual [chuckle]".[13] They were now dealing with a powerful case of cognitive dissonance. They had expected to see the runway much earlier. At this point, they were still following the vertical speed command of 1,500 FPM down. Not seeing the runway at this very late point constitutes a significant failure, which really takes us back to the approach briefing, because to recognize this as a failure, you have to have briefed the expected ceiling height. In my case at Monterrey, I had the advantage of a procedure, unique to the MD-80 with no VNAV and long since gone by the wayside, of refraining from setting the missed approach altitude until we had the runway in sight. The UPS crew, like everyone else using either published decision altitudes or derived decision altitudes, had no such procedure; they had already set their missed approach altitude, so the autopilot had nothing to capture at the MDA.

The Monterrey story gets more interesting. We had no fuel for a second attempt, nor did we see any point since we had seen nothing, so we diverted back up to McAllen, Texas, and refueled. Returning to Monterrey, the strong east wind had subsided, and the first officer suggested we fly the ILS to the opposite direction runway with an acceptable tailwind. We set that up and were vectored onto final approach, with a fairly sharp interception angle. The localizer swung across, the FMA indicated it was captured, and the autopilot happily flew through the localizer on the assigned heading and kept going. We abandoned the approach, got vectors for a second attempt with a much shallower intercept, and the same thing happened.

The second time, frustrated, I briefly contemplated disconnecting the autopilot and manually tracking the ILS. This would have been easy to do; the localizer and glide slope indications appeared valid. But a preoccupation with failure dominated. Something was fishy; we had already concluded that the weather observation was faulty. We had joked that the fellow issuing the weather report … still four hundred scattered, eight hundred broken … was probably camped on his balcony downtown, where the refinery gas flares had burned off the stratus layer quite nicely, with a martini in his hand, and not out at the airport. I really didn't know whether our equipment was screwed up, whether the ILS ground equipment was truly monitored, or what might be going on. Both conclusions were derived from a sensitivity to operations. This was clearly a case in which we were encountering things which "disconfirm, are unpleasant, feel uncertain, seem possible, are implicit, and are contested". The mistake I did not want to make was to descend toward an airport just short of 13,000-foot mountains without knowing why my autopilot would not capture the localizer, particularly when the FMA claimed that it had.

It turned out that the first officer, a great guy but very new to the airline and airplane, had failed to set the localizer inbound courses properly. Third time's a charm; everything performed as it should, and we broke out in moderate rain at three hundred feet. Had we attempted the earlier non-precision approach, we would have never seen the runway, again, as in the first approach.

I don't particularly see that night as one of the more stellar moments of my career. I should have seen the localizer course error; there were two parts to that checklist item … checking both localizer courses … and my poor first officer missed them both. I corrected his localizer course, but, satisfied with my sharp awareness, missed my own, which, since I was the pilot flying using my autopilot, was the one that really mattered. To be fair to the first officer, this was a very dynamic situation for which the Advanced Qualification Program, and LOFT, had not prepared him. He was working his ass off trying to do whatever I needed. The Mexican controller, amused, rather wittily pointed out we could keep trying as long as I paid for the coffee.

All of this is easy to armchair. In the Cali and Birmingham cases, the crews did not appear to employ either a preoccupation with failure or a sensitivity to operations, nor did they engage circumspection. Both were fatigued to one degree or another and both were ready to get on the ground. They were both caught squarely by action-based cognitive dissonance. On the other hand, both cases were routine situations that all of us have encountered many times. In constructing work-arounds to get the job done, our past experience drives the normalization of deviance, which facilitates simplification and spares us the effort of consciously altering the dominant thought.

A different outcome can be seen in this report from the NASA ASRS:

> We were at 2,600 FT and cleared for the approach. The Captain was pilot flying and I was pilot not flying. We were a little fast and trying to slow and configure. We captured the localizer. While tracking the localizer inbound, I performed a landing check to the line, checked in with the Tower, and set the missed approach heading and altitude. When I looked up the Captain had the airplane in VS mode descending at 1,300 FPM. I checked the glide slope and it had not come "alive" at all on both of our PFD's. I immediately stated "Go Around". Simultaneously, ATC said "low altitude alert, check altitude". We got the "glide slope" aural warning and also "terrain, terrain" aural GPWS warning. We executed a go around and diverted. We were at or just below 1,000 FT AGL when the go around was initiated. The threats were the mountainous terrain at night in icing conditions down to minimums. Also, we were on our 6th leg of the day. Errors were the pilot flying, the Captain, began a descent without a glide slope at 1,300 FPM. I will be extra vigilant when operating in these types of conditions, as I feel I already pretty much was.[14]

NOTES

1. Karl Weick, Kathleen Sutcliffe, *Managing the Unexpected: Assuring High Performance in an Age of Complexity*, Wiley. Kindle Edition, 2015, 90.
2. Thomas Aquinas, *Summa Theologica*, translated by Fathers of the English Dominican Province, e-artnow Kindle Edition, 2013, IIa-IIae, a.5, ad.obj.2.
3. Aeronautica Civil of The Republic of Colombia, Santafe De Bogota, D.C. – Colombia, Aircraft Accident Report, Controlled Flight Into Terrain, American Airlines Flight 965, Boeing 757-223, N651AA, Near Cali, Colombia, December 20, 1995 (translated by Peter Ladkin, Universität Bielefeld, Germany)
4. Ibid, 33.
5. *Summa Theologica*, q.49, a.7, ad.obj.3.

6. Gene Rochlin, *Trapped in the Net: The Unanticipated Consequences of Computerization,* Princeton University Press, 1998, 109.
7. Karl Weick, Kathleen Sutcliffe, "Managing The Unexpected", 2nd Managing the Unexpected in Wildland Prescribed Fire Workshop, Jacksonville, Florida, February 28, 2005.
8. Ibid.
9. Aeronautica Civil of the Republic of Colombia, 32.
10. Eddie Harmon-Jones, David M. Amodio, Cindy Harmon-Jones, "Action-Based Model of Dissonance: A Review, Integration, and Expansion of Conceptions of Cognitive Conflict", in Mark P. Zanna Editor, *Advances in Experimental Social Psychology*, Vol. 41, Academic Press, 2009, 128.
11. National Transportation Safety Board, AAR-14/02, Crash During a Nighttime Non-precision Instrument Approach to Landing UPS Flight 1354 Airbus A300-600, N155UP Birmingham, Alabama August 14, 2013.
12. Karl Weick, Kathleen Sutcliffe, *Managing the Unexpected.*
13. National Transportation Safety Board, AAR-14/02, 147.
14. NASA Aviation Safety Reporting System, Report Number 993408, February 2012.

11 Strategy over Program

Edgar Morin has argued that the correct approach to the ecology of action lies with strategy.[1] He has argued that short-term programs become stuck in modified or unstable environments, and that strategy must prevail. So how do we build a strategy? For that matter, what is a strategy? A purely technical definition defines it as a high-level plan for to achieve one or more goals under conditions of uncertainty. But the best explanation I've seen was given by Sir Winston Churchill during his 1946 "Sinews of Peace" speech at Westminster College, the same one in which he coined the term "iron curtain". Churchill said that

> When American military men approach some serious situation, they are wont to write at the head of their directive the words "over-all strategic concept." There is wisdom in this, as it leads to clarity of thought.[2]

That is precisely what we are aiming at ... an overall strategic concept that leads to clarity of thought. How do we define and construct such a strategy?

Over the course of aviation history, there have been numerous attempts at defining a universal strategy for flight; the nine axioms that precede the tenth axiom are one example of a roadmap, a set of guiding principles, which could, perhaps, define a type of high level plan for achieving one or more goals under conditions of uncertainty. For example:

- The competent and properly trained pilot will always fly by his judgment as developed through experience, and not allow extraneous sensations to warp his judgment. (Instrument flying, of course, is another matter.)
- No competent or properly trained pilot will ever approach a stall or spin during the execution of any normal maneuver.
- No careful, well-trained, competent pilot will ever be caught in a position from which a reasonably safe emergency landing cannot be made.
- A fuel supply check before each take-off will insure an ample supply for safety.
- Safety and altitude go hand in hand, particularly during periods of emergency.[3]

But, as with so much of the era's admonitions, the only real clarity of thought from these would be that once you had inadvertently stalled the airplane, you would be clear in the knowledge that you shouldn't have. Thanks for that.

Following World War II, given the growing awareness of our capable and competent airman's carefree attitude and lack of interest in abstract thought, subsequent efforts to impart strategy in the cockpit have often been expressed as coded mnemonics and slogans, such as the classic "aviate, navigate, and communicate". Aeronautical decision-making has become rife with acronyms like the 5 Ps 3 Ps, PAVE, TEAM, CARE, and DECIDE.[4] Perhaps one of the best of these ideas is

DOI: 10.1201/9781003369677-11

known as FOR-DEC and was developed by the folks at Lufthansa.[5] This stands for Facts, Options, Risks and benefits, then a hyphen intended to kind of shift cognitive gears, followed by Decision, Execution, and Check. If we look more closely at the idea, we can see that facts and options are really St. Thomas' ideas of taking counsel and judging of what you have learned. Risks and benefits are nothing more than Morin's "awareness of wager". Decision and execution are the command element of prudence, and the final item, check, is simply circumspection, looking to see whether the plan as envisioned is still meaningful to the situation.

The most recent and developed concept of a universal strategy is certainly threat and error management. At its core, this approach is aimed at Morin's basic idea within the ecology of action, the notion that an unruly trajectory, the threat or the error, must be torpedoed. Sadly, threat and error management is almost always presented in terms of specific risk, identified threats, threats that *should* be identified, errors that *should* be actively trapped ... lions ... essentially focusing on extensive lists of example threats, errors, and undesired aircraft states, which creates a mental model of turning the known-unknowns into known-knowns. This concept is essential; as we discussed in Chapter 4, all zebras, regardless of stripes, pay very close attention to the movement of the grass and the sniff of the air. However, spotting the lion in the grass is not enough; a specific risk approach can easily default to the capable and competent aviator never allowing a crack-up, that is, the capable and competent aviator will always identify the threat and then manage it, which is the inescapable pull of the word "management" in threat and error management. Threat and error management approached from a set of lists does not truly answer the ecology of action.

In fact, the mnemonics are really not strategies at all, but tools useful when dealing with specific risk. Threat and error management, in isolation, is also a tool, but not really a strategy. None of these ideas define an "over-all strategic concept", to use Churchill's words. To build a truly universal strategy, we need to step back and look at the problem from one level higher, that of probabilistic risk. We need an approach to the ones you don't see coming, the known-unknowns that never become known-knowns, as well as the unknown-unknowns, the black swans.

In recent years, concepts, tools, and structures that focus on the management of specific risk have collectively become known as Safety One. This term refers to traditional ways of investigating and managing safety, which is almost wholly centered on the prevention of specific accidents.

Safety One is heavily invested in human error and exists for all of the reasons we have discussed throughout the book, in particular, the idea of risk aversion, the dependence on the narrative mode of thought, and Taleb's description of the instinctive drive to explain causality following something we didn't see coming, something nothing in our experience would have anticipated, a black swan event.

Figure 11.1 illustrates a model of how this idea works. We looked at this figure in Chapter 1, when discussing Rene Amalberti's ideas of violation behavior; however, the diagram was first developed by Jens Rasmussen.[6] The initial space of safe operation is bounded by economic forces on top; any drift toward that line reduces profitability. On the bottom, workload as well as social and individual forces all push upward along the axis of personal gain. In other words, our natural tendency is to drift farther upward, toward less workload. The left is bounded by rules, procedures,

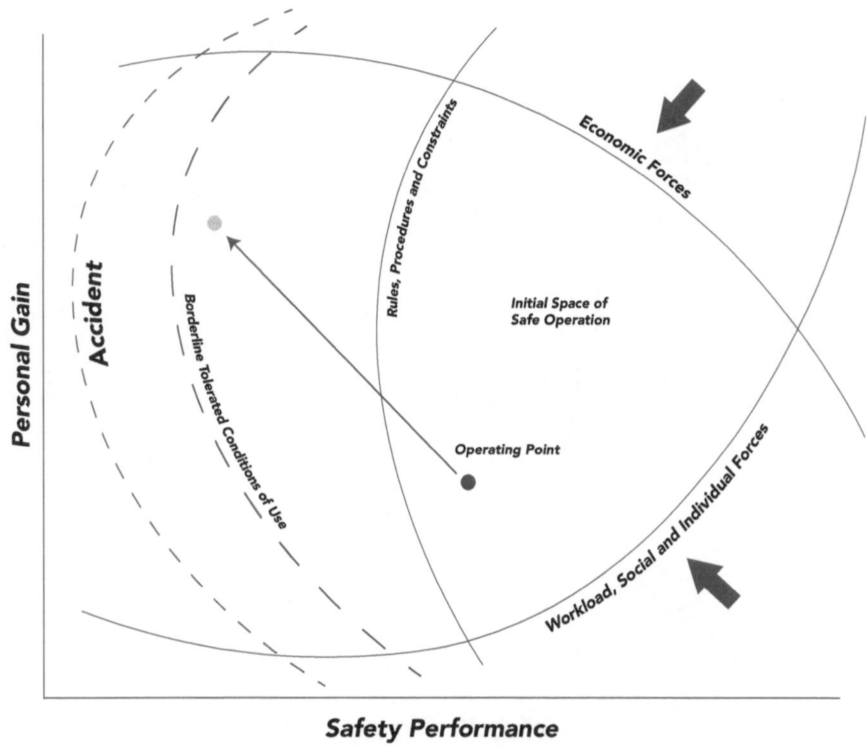

FIGURE 11.1 The space of safe operation, with Rasmussen's model of drift toward failure and Amalberti's model of violation behavior.

and constraints. These form a guardrail designed to steer the operation well clear of the far-left region, which is the domain of accidents.

The general idea is that due to our drift toward easier workload, including routine and optimizing violations, we often cross to the left of the rules and procedures line, into an area that you will recall Amalberti described as the borderline tolerated conditions of use (BTCU).[7] Within this region, the margins of safety have been eroded and the likelihood of an accident is considerably greater.

The classical Safety One response to this drift is the addition of more rules, procedures, and constraints. More recently, the general education of the workforce in safety culture augments the addition of more procedures, and together they push the boundary of the borderline conditions back toward the safety of rules and procedures. This is shown in Figure 11.2, with the arrows representing a safety culture pushing the borderline tolerated conditions back to the right, much closer to the rules and procedures line.

A different approach has been argued by Erik Hollnagel, which has become known as Safety Two.[8] From the perspective of Safety Two, the focus is shifted from studying what went wrong to studying what went right. Hollnagel's argument is that accidents are characterized by the absence of safety, ergo the study of accidents in isolation does not yield much knowledge about what safety is, but a great deal of

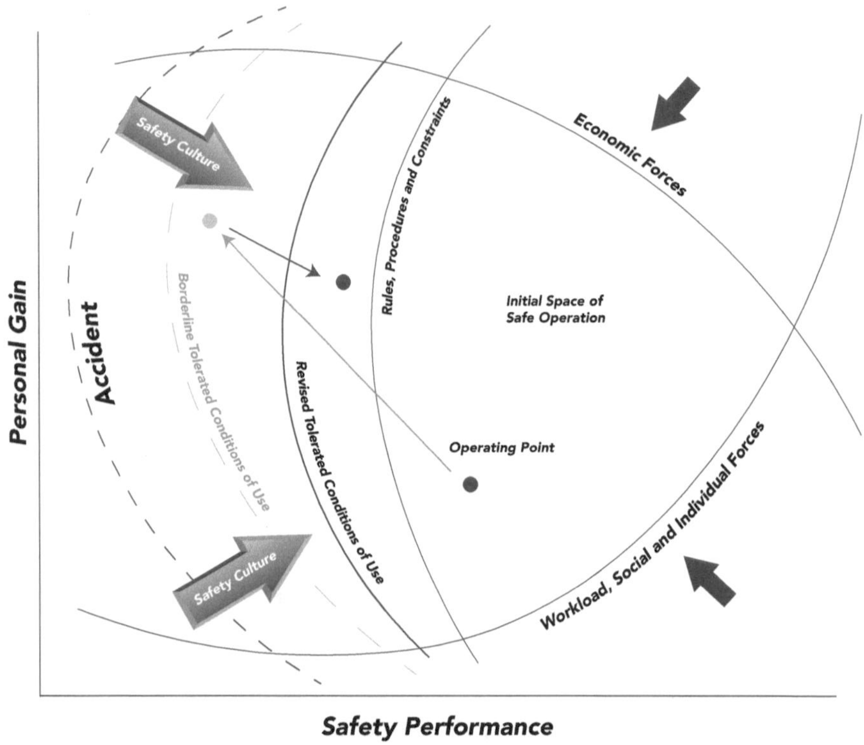

FIGURE 11.2 The Safety One response to drift.

knowledge about what safety is not. He makes an amusing point by suggesting that we do not learn much about a happy marriage by studying divorce.

The argument for Safety Two also posits a significant paradigm shift from Safety One. Whereas the Safety One approach looks for the peripeteia, the departure from normal, the failure, Safety Two argues that things go wrong in essentially the same way that things go right. Safety Two philosophy identifies variability in human performance and recognizes this variability as a necessary feature of safety. The variability in human performance is largely the result of the differences between work-as-imagined … proceduralization … and work-as-done, which incorporates shortcuts, work-arounds, and other informal bridges across procedure. This is a critical acknowledgement of the small but continuous human interventions that are necessary to accomplish the objective of a safe outcome. In parallel with this, Safety Two philosophy recognizes complexity and emergence. In this approach, the accident results from an emergent behavior of the system, in which the trajectory of the accident is not linear, and it cannot be broken down into components, exactly as we discussed in Chapter 5. The emergence is a higher-level outcome of lower-level variability in human performance, and I would also argue the variability on different levels in the operating environment, generating local feedbacks and relationship changes that lead to further variability, further feedback, action endogeneity, and ultimately yield an unpredictable outcome.

In other words, you will never see the one coming that gets you.

The upshot is that Safety Two philosophy dispenses with the narrative mode of thought as a means of understanding accident causality. What we are instead interested in is understanding the environment as an open system that is sensitive to initial conditions, very much subject to complexity and emergence, then understanding the significance of human variability in achieving a safe outcome, and finally being cognizant of the conditions under which it may be difficult or perhaps even impossible to monitor and control performance variability.[9]

This is precisely the point at which Morin's ecology of action becomes critical. Work-as-imagined, or proceduralization, is incapable of comprehending the sphere of interactions at the moment, and it will not anticipate an action escaping from the designed intention. Work-as-imagined is a program, and it can become stuck. To get from there to the safe outcome, the human pilot must take into account complexity, "random incidents, chance, initiative, decision, the unexpected, the unforeseen, and awareness of deviations and transformations".[10] We must follow our actions, identify unruly trajectories, and shoot them down.

In Figure 11.3, I have modified the original picture to show some of the more applicable aspects of Safety Two. The left axis is no longer focused on personal gain,

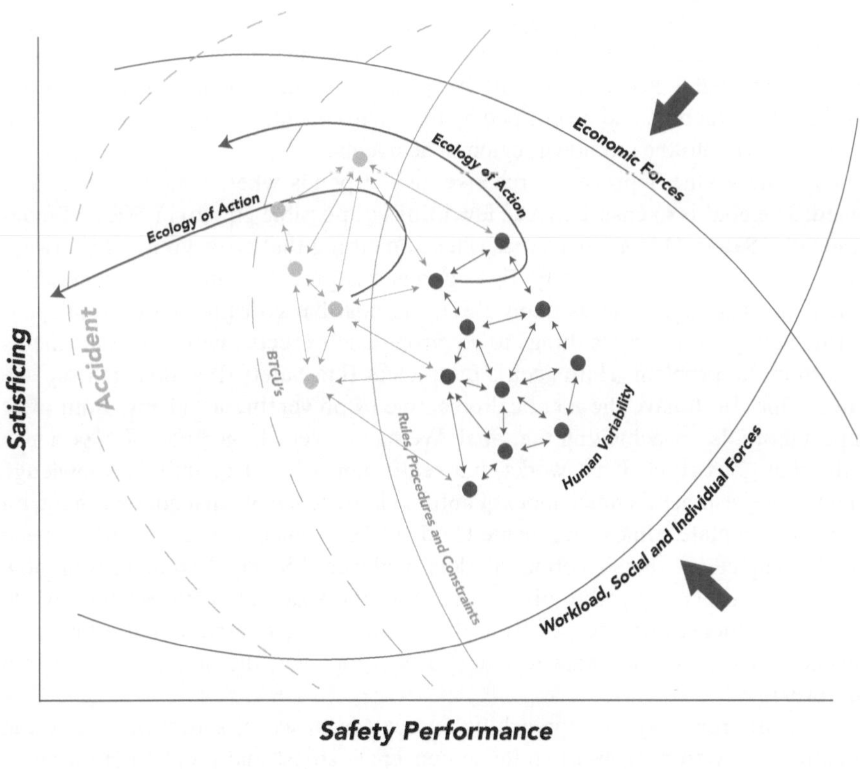

FIGURE 11.3 The space of safe operation, with the safety two model of human variability, illustrating the differences between work-as-imagined and work-as-done.

but rather a concept first developed by Herbert Simon, which he called "satisficing".[11] The basic idea is that we will work to achieve a balance between the quality of the work output and the effort we are inclined to put into the work. We all know that some pilots put considerable more effort into the job than others; some, like me, are obsessive. These are different degrees of satisficing. In the region of lesser economic performance pressure, there is more room for satisficing; in other words, the lower the economic pressure, the less work we may apply because we have the space to be satisfied with a lower quality output.

Instead of one point of operation, I have depicted the actual state of human variability as multiple points of operations. This variability is strongly influenced by local feedback, which is represented by the bidirectional arrows. The work-as-imagined lies to the right of the rules, procedures, and constraints boundary. Most of the time, our own performance variability remains within this region. However, the real world often intrudes on work-as-imagined, and we occasionally need to cross the line, departing the initial space of safe operation and venturing into the BTCUs. This is the difference between work-as-imagined and work-as-done. Some of this may be due to satisficing, but some of it may be due to the realities of the economic pressures of efficiency or simply doing what is necessary to achieve the assigned task. As the economic pressure builds, the points of operation descend toward increasing workload.

At any point, whether we are operating within the space of safe operation or to the left of it, within the BTCUs, the ecology of action is engaged. Actions we may take, initially expected to lead back toward the space of safe operation, may enter Morin's spheres of interaction, and be grasped by the environment, driving the outcome well to the left, out into the unknown region of accidents.

The Safety One approach is reactive and responds when something has happened. The goal is to ensure that as few things as possible go wrong. Viewed independently, Safety One assumes that when something that goes wrong, it is always a potential accident. The Safety Two approach is proactive and anticipatory. The goal is to make sure that as many things as possible go right. Viewed independently, Safety Two expects things to go wrong and expects intelligent adjustments to correct the problem. Hollnagel is quite clear that Safety Two does not replace Safety One. Both have the identical objective of preventing accidents. Both offer important tools for achieving that goal. We have a very large body of experience with what goes right, what works, but really not a lot of quantified knowledge about it. We also have a much more quantified knowledge of what goes wrong, with a far less complete, almost fragmented body of experience, with considerable open space occupied by Taleb's notion of silent evidence.[12] Safety Two focuses on how we use variability to capture divergence from the space of safe operation, while Safety One focuses on escapes from the space of safe operation and establishes mistakes that we do not want to make. Taken together, the output is a dynamic non-event.

So, a different way of approaching strategy is to write, across the top of the directive, the words "over-all strategic concept – adjust and protect the margins, arm the constraints, and engage variability so as to remain the master of the aircraft". Most of us have been doing this for our entire careers, creating extensive lists of Weick's dynamic non-events, while our safety education has largely centered on Safety One, a much shorter list of dynamic *events*. It is worth unpacking just what is going on here.

THE MARGINS

The first element of such a strategy is to anticipate variability in human performance, anticipate the variability in the environment, anticipate the ecology of action, and focus on strengthening and protecting the margins. The margins are necessary to tolerate the variability in the environment, the probability of the environment becoming modified, as well as provide space for variability in human performance, and finally to enable that performance to identify and shoot down unruly trajectories.

To construct a strategy, the first step is to identify critical margins, and particularly weak spots in those margins. SOPs are constructed largely with margins in mind, but SOPs are programs that can become stuck. Whereas Amalberti's original idea of an initial space of safe operation was defined by procedures, it is critical to recognize that the safe operating space is actually defined by features external to SOPs, features that include the margins but also include mistakes we don't want to make, a reluctance to simplify and a sensitivity to operations. On a good day, most SOPs will keep you within a margin-specified safe operating space very nicely, but it is through the continuous loop of mindfulness … a rich awareness of discriminatory detail … that we identify a modified environment, the onset of instability in the environment, and then actively transition the centerline of the safe operating space, re-route it, from one procedure to another, and quite possibly leverage human performance to construct a modified procedure entirely. Eventually, we may complete the flight without a procedure at all, but still within a safe operating space with the margins well-protected.

Consequently, a proactive and anticipatory strategy begins with a comprehensive knowledge of the margins and a commitment to protect them. The margins are never incorporated into the strategy; the protection of the margins is the cornerstone of the strategy. It is difficult to overstate this point as it represents a paradigm shift from the ubiquitous drift away from the initial space of safe operation toward the BTCUs, where efficiency may be optimized and cost minimized while still retaining a modicum of a margin. Margins do not exist for optimization, or cost manipulation, or as any kind of a backup plan. Margins are never engaged as part of the plan; they exist solely to manage an open system and the ecology of action.

Let me give an example. It is fairly common for scheduled airlines to put a high premium on closing the main cabin door on time. Regardless of what time we were supposed to close the door, I would not allow the main cabin door to be closed if any other cabin door was still open, that is, the caterers were still loading the aft galley, or if the aft cargo door was still open, or if I was aware of any other known delay of significance in starting the pushback, such as shift change on the ramp, etc. The reason I did this was to protect the certification margins for evacuation. Unless all doors except the forward cargo door were closed and no delays anticipated, I had no idea how long we might still have to sit on the gate (the forward cargo door was typically closed only after the jetway was pulled to accommodate the last of the gate checked bags). If the cabin suddenly started filling up with smoke, and we had to evacuate, I needed all exits available. If the main cabin door was closed, the agent would pull the jetway back ... but not far enough to deploy the slide from that door. So one exit is already blocked, and we haven't even started the evacuation.

Discussing this with a chief pilot one day, she pointed out that the evacuation capability is certified with 50% of the exits blocked ... so why did I care about one being blocked for a brief period? Well ... because the certification with half of the exits blocked is intended to provide a margin against exits failing during the evacuation *for reasons that cannot be foreseen*. It is *not* intended to be used for operational optimization. With the main cabin door still open and the jetway still attached, a high-volume exit is available, the flight attendants have an emergency deplaning procedure that uses that exit, and, if need be, they can still arm and deploy the remaining exit slides. Once it is closed, it is blocked by the jetway (in most cases, although sometimes not, which changes the strategy).

So is this margin established through procedure or regulation? No. FAR 121.570 says the following:

> Each certificate holder shall ensure that, at all times passengers are on board prior to airplane movement on the surface, at least one floor-level exit provides for the egress of passengers through normal or emergency means.[13]

This essentially says that before moving the aircraft, with a full load of passengers, only one of the four (Boeing 737) floor-level exits needs to be available. This is an example of a rule that prescribes absolutely no margin whatsoever; in fact, it destroys the certification margin. Essentially, if we use Amalberti's definitions, this rule has codified the BTCUs as part of the initial space of safe operation.

Before going further, we need to explore margins in more detail and consider how to prioritize them. Not surprisingly, it turns out that margins exist at several levels within the operating system and environment. We are quite literally surrounded by them.

It is common for a bank angle limit to exist between V_2 and V_2 plus 15 or 20 knots; this provides a margin above a stall due to increased G during initial acceleration. Cruise speeds generally provide a margin above stall at bank angles up to thirty degrees, typically with a fifteen degrees margin over that. A wind additive provides a margin for gusts above V_{ref}. V_{mo}/M_{mo} provides a margin to protect the structural integrity of the airframe. These are what we might classify as aerodynamic margins; angle of attack, bank angle, G loading, and thrust or power. Since keeping the airplane actually airborne is the critical idea, the aerodynamic margins are one of two predominant margin sets that must be prioritized above all else. When we say fly the airplane first, or aviate, this is exactly what we mean. As master of the aircraft, you must use the flight controls and power controls to assertively protect aerodynamic margins.

Minimum altitudes of all sizes and shapes ... MEAs, MOCAs, grid MORAs, MDAs, DHs, and the like ... are all designed to protect an altitude margin. Driftdown requirements in FAR 121.191 require a positive slope of at least 1,000 feet above any terrain or obstacle within a five-mile margin on either side of the route, or, if a positive slope cannot be achieved, a 2,000-foot margin between the airplane and any terrain or obstacle within a five-mile margin on either side of the route. These are altitude margins, the second predominant margin set. They are of more or less equal importance with the aerodynamic margins, depending on the circumstance. Suffice to say that if the priority between the two is open to discussion, you're already in deep yogurt. For example, a normal approach constitutes the planned diminishment of the altitude margin while carefully protecting the aerodynamic margins. On the other hand, the windshear escape maneuver prioritizes the altitude margin while squeezing the aerodynamic margins to almost nothing … the stick shaker may be triggered intermittently while extracting maximum performance. The protection of aerodynamic and altitude margins is the point of the old, original steam gauge situational awareness … where are we, where is the ground, where are we going, and at what angle of attack?

On December 17, 1987, AVAir 3585 experienced a dual flameout while operating a Fairchild Metroliner during a night visual approach at Washington Dulles.[14] The crew had failed to operate the engine anti-ice, and upon entering warmer air, ingested ice into both engines. The landing gear and flaps were already extended. The captain opted for an off-airport landing in a farmfield. At approximately two hundred feet, while aiming at a lighter patch in the middle of all the dark patches, he sensed an aerodynamic stall developing. He shoved the controls fully forward to break the stall, then pulled the yoke back into his gut to break the descent just prior to touchdown. There were a couple of injuries, the airplane was destroyed, but no one was killed. In this case, the captain had correctly prioritized protection of the aerodynamic margins, in the dark with no power, retaining the control he needed to flare. I knew the captain and investigated the accident. I've spent the past several decades wondering whether I would have had the same presence of mind.

At the next level, FAR 121.195 requires that the aircraft be dispatched at a weight that, based on normal fuel consumption, will allow it to stop within 60% of the available runway length on the most favorable landing runway. FAR 121.189 requires that the takeoff weight must assure clearance of all obstacles within two hundred feet of the departure flight path by thirty-five feet. These are performance margins. They are probably the margins we are most familiar with, as they demand regular calculation prior to takeoff and landing. Note that in cases where you are not runway or climb limited, these margins are already fattened up. And as we discussed in Chapter 4, the reductions in landing margins implemented through the TALPA program raise serious concerns about the adequacy of those margins; 15% is never going to be realistically adequate. On the other hand, if you are landing on a 12,000-foot runway and only need 5,000 feet at your particular landing weight, you have a very cushy margin. In any case, the whole point behind company policies like landing within the first 3,000 feet or one-third of the landing distance available is to protect these margins.

Fuel reserves are a performance margin. What altitude is the reserve fuel predicated on? What altitude and speed are planned for the diversion to the filed alternate? What cruise altitude and mach is the initial arrival fuel based on? Which exemption from the regulatory requirements are we operating under? 10% of the total flight time or 10% of the time in Class II navigation? 5% of the time in Class II navigation? A tailored reserve plus thirty minutes at 1,500 feet? These are common margins that we work with every day, but it is worth understanding how these margins are arrived at, in order to evaluate them for weak spots.

At a different level, single generator operation requires an immediate landing at the nearest suitable airport, in order to protect the thirty minutes or greater margin of electrical power available from the battery. Operation with a single pack is often restricted by the MEL to FL 250, in order to improve the margin of safety should the remaining pack fail. Non-ETOPs operations require that a suitable landing airport is always within sixty minutes flying time, in order to provide a margin against the probabilities of additional system failures. With those probabilities reduced through design and certification, ETOPs programs may increase that time, while still retaining a suitable margin.

These types of margins can be classified as airworthiness margins. While the aerodynamic and altitude margins must prevail … airworthiness is of little importance if the airplane is not actually flying … the airworthiness margins are of subordinate yet quite proximate importance. Airworthiness margins are less visible than aerodynamic, altitude, or performance margins. It is essential to recognize that the airplane was certificated to a standard of safety that presumes predicted and defined airworthy configurations, ranging from fully functional through several stages of degradation. If the airworthiness begins to degrade, the entire point of the MEL, on the ground, and the QRH, in the air, is to capture the degradation and secure the configuration, thus protecting whatever functionality margins were built into that configuration. To wit: the prop needs to be feathered, the fire needs to be extinguished, the runaway stabilizer needs to be shut down, and the cabin altitude needs to sustain life. And so on. Here resides the idea of ensuring that all emergency exits are available once the main cabin door is closed.

Note that there is a significant interdependency between airworthiness margins and performance margins. The manufacturers provide detailed landing distance assessment data for a wide range of degraded airworthiness conditions. En route fuel burns are completely dependent on configuration. That data, and the margins it protects, cannot be properly applied if the configuration of the aircraft does not meet the applicable airworthiness criteria, i.e., the single engine performance is all predicated on the timely feathering of the prop.

Beyond these tangible margins, recall Morin's point regarding the modification of the environment and the instability that can result. Within the cockpit, one of the most important margins that must be protected is that of Stanley Roscoe's idea of residual attention. Cockpit workload must be managed to protect this margin; adequate residual attention must be retained for circumspection, to identify stuck programs, and to identify unruly trajectories resulting from actions taken, and shoot them down. It is here that SOPs can play a significant role. However, caution is warranted in the event that any procedure, as a short-term program sequence, becomes stuck. The margin of residual attention must always take priority over the procedure. When the procedure does become stuck, stop, step away from the procedure, ensure that the primary aerodynamic and altitude margins are protected, and take some time to expand the margin of residual attention. Then resume the procedure as appropriate. In most cases, the margin of residual attention can be protected by workload distribution and task assignments. Note that one of the most effective ways of protecting this margin, and fattening it as needed, is to simply slow the airplane down.

There have been numerous accidents that can be traced to an erosion in the margins of residual attention. This is exactly what happens when pilots become pre-occupied with the FMC or automation in general. The most well-known residual attention accident was undoubtedly Eastern 401, a Lockheed L-1011 that crashed in the Florida Everglades on December 29, 1972.[15] The crew had entered a holding pattern at 2,000 feet in order to resolve a landing gear indication problem; the nose gear light was burned out. The flight engineer went down into the E&E compartment below the cockpit to visually confirm gear extension. The pilots were apparently occupied with the burned-out light, failed to notice that the autopilot had transitioned into control-wheel steering mode, and also failed to note the altitude alert chime when the airplane had begun a slow descent. The last thing they heard was the radio altimeter counting down the last few feet.

Figure 11.4 shows the same space of safe operation shown in Figure 11.3, but with a curve added to represent the margins. This curve is a kind of aggregate

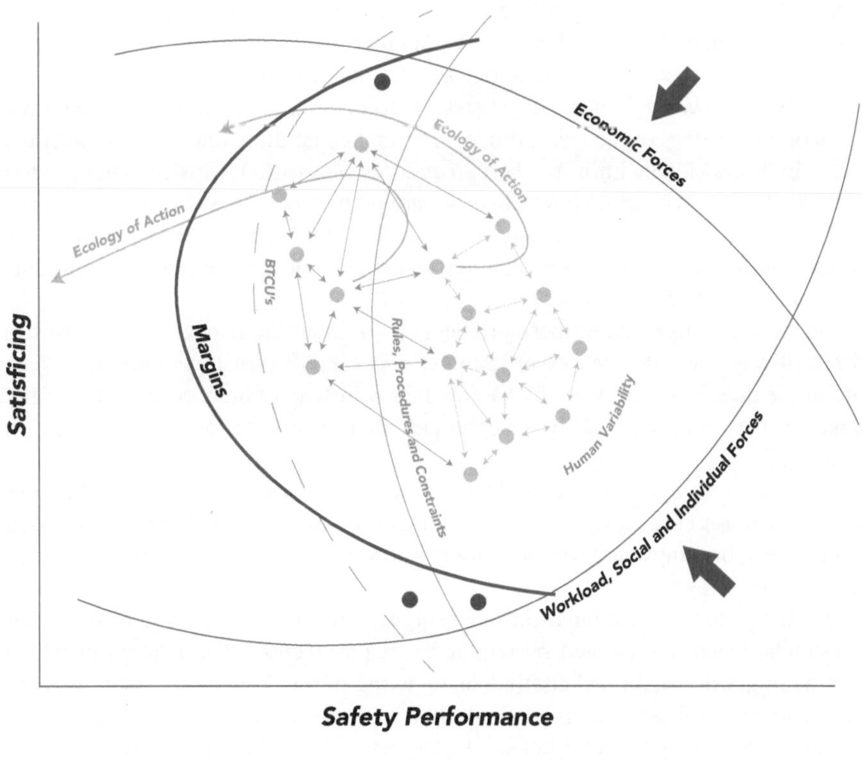

FIGURE 11.4 The space of safe operation with margin boundary.

margin line; it includes operational margins of aerodynamics, altitude, airworthiness, and performance, but it is also weighted for the margin of residual attention against increased workload. There are three things to take note of. First, generally speaking, the margins we operate with lie to the left of the rules, procedures, and constraints, and pretty much to the left of the BTCUs. However, as we move toward greater degrees of satisficing, less work, and less concern for profitability, the BTCU boundary drifts well left of procedures and somewhere up there crosses the margin boundary.

Second, as we descend the satisficing axis in response to increasing efficiency and profitability, we increase workload and press against stronger social and individual behavioral forces. At some point, the margin of residual attention is pretty much gone; the aggregate margin boundary will literally cross over the rules, procedures, and constraints curve in the vicinity of maximum sustainable workload.

These first two observations lead to the possibility of three operating points, represented by black dots. In the upper example, the large gap between rules, procedures, and constraints and the BTCUs allows a drift to the left that approaches operational margins; this is a region of work-as-done that reflects poor efficiency and little workload.

The lower left black dot also indicates a drift into the work-as-done region while operating under very high expectation of efficiency and profitability, with the consummate high workload. Any residual attention has disappeared, and we are now outside of the aggregate margin line since, without any residual attention, we really have no idea where we are with respect to any margin … and since the margin actually bounds the space of safe operation, we are outside of that.

Finally, the lower right black dot shows a similar condition, but in this case within the theoretical safety of the rules, procedures, and constraints line. This is the effect of very high workload within standard procedures. The residual attention margin has evaporated, and we cannot know where we are with respect to any of the margins.

The third observation from Figure 11.4 is that the aggregate margin boundary creates a buffer that is more likely to allow an unruly trajectory to be shot down. This is the whole point of margins.

The strategic approach to these problems is to fatten the margins. We do this in dozens of ways; we add fuel, we elect to use max takeoff thrust, we fly a wider path around the thunderstorm, we select autobrakes 3 instead of autobrakes 2. Carefully managed, these decisions shift the aggregate margin line to the left, as shown in Figure 11.5.

With conscious attention, the problem of the lower black dots is resolved, because we have increased the margin of residual attention, through such things as workload management, briefings, and prioritizations. And we have even more buffer to capture unruly trajectories.

Adjusting the aggregate margin line requires careful consideration of the original, standard margins created by regulatory requirements, identification of which margins might be weak, and continuous updating of the strategy to ensure that the margins remain robust. This is particularly true of the margin of residual attention; any particular unruly trajectory or stuck program will eat that up in a heartbeat.

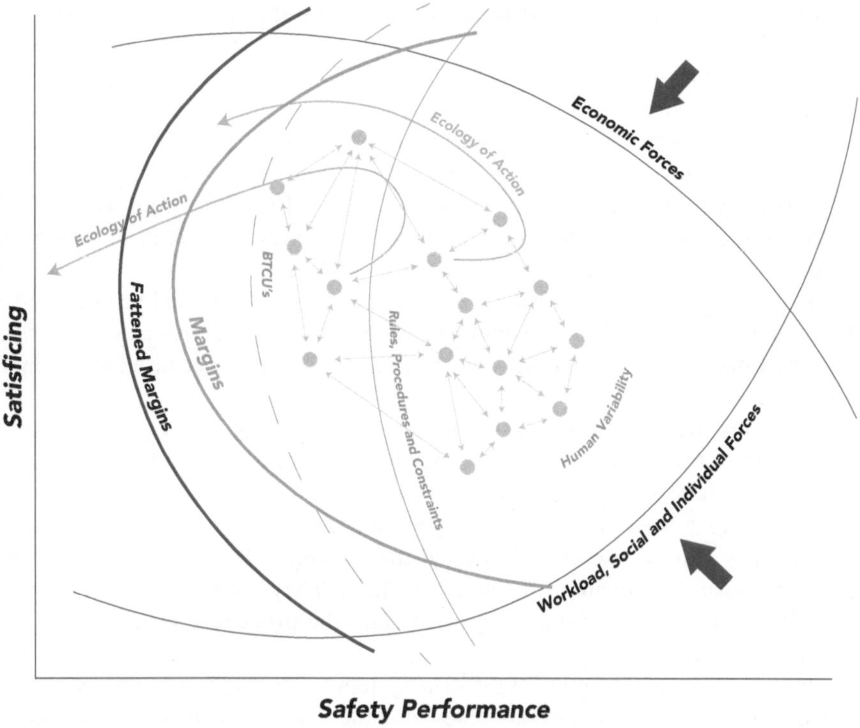

FIGURE 11.5 The space of safe operation with increased margins.

THE CONSTRAINTS

Tangent to the protection of the margins is the protection of constraints. This is the second element in the construction of a strategy. Recall Amalberti's idea of constraints and fail-safe procedures designed as defenses against error.[16] They key word here is constrain; this the second element in building the management of variability in human performance into strategy. The key to monitoring and controlling variability in human performance is to protect the constraints.

The term constraint is used by the system engineering people in the design of large, complex technical and techno-social systems. A constraint is type of structure, physical or procedural, that is placed specifically at one level to prevent systemic interactions that may lead to emergent phenomena at a higher level.

A very simple example of a constraint is the knuckle-buster on the 757 cabin door. When you arm the door, a rounded wedge extends in a manner that interferes with your hand if you try to open the door handle. The goal is to preclude you from inadvertently opening the door with the slide armed. If you have to open it during an evacuation, your knuckles will hit the wedge, compress a spring, and it will retract slightly as your hand passes over it. If you are not evacuating, presumably you will stop rotating the handle when your knuckles bump the wedge and disarm the door like you were supposed to. Another unruly trajectory successfully torpedoed.

It turns out that we are surrounded by constraints. The gear warning horn is a constraint. The overspeed clacker is a constraint. A thrust lever interlock is a constraint. And more importantly, we are surrounded by procedural constraints built into the SOPs. The captain's signature is required on the flight plan; this is a constraint to ensure that he or she has reviewed it. The "point-and-shoot" method of resetting the altitude alert on the mode control panel requires that both pilots physically point to the new altitude and vocalize it; this, too, is a constraint intended to preclude altitude errors. The final check of takeoff fuel prior to brake release is a constraint. The procedure for disconnecting a constant speed drive has a confirmation step ... another constraint ... built into it, because the action is irreversible in flight.

Constraints are built into SOPs. The assumption used by the system safety engineers is that you will execute the procedure correctly and in a timely fashion. If they think there is a good chance you're going to screw it up, they'll install something like a knuckle-buster or a gear warning horn to focus your attention. But they can't chase you all around the cockpit with more and more constraints because you are sloppy or non-compliant. They are going to assume that you will protect the constraints they wrote into the procedure by executing the procedure correctly.

Indeed, there is a fair argument that true system safety engineering has yet to be applied comprehensively to aircraft design and the design of operating procedures. Some constraints may not be as consistently successful as they should be. Some may result in unintended consequences, which itself is a serious threat because it quickly leads to informal procedures that short-circuit the constraint. But a large part of tracking the centerline of the safe operating space is a disciplined execution of a SOP to make sure the constraints work like they are supposed to. These constraints are specifically designed to prevent actions from being grasped by the environment and entering unruly trajectories. They simply do not work if the procedure is executed haphazardly or not executed at all.

And that is precisely what happened in the Air Transat maintenance hangar; two procedural constraints were in place ... documenting the engine configuration upon acceptance and reading the actual service bulletin when doing the engine change ... and neither constraint functioned as intended. In the latter case, the environment was modified by the failure of the company computer network, making access to the service bulletin difficult. That procedure became stuck, so a workaround was devised. This was a significant drift away from the initial space of safe operation and clearly crossed beyond the BTCUs.

Which is also precisely what happened in the Air Transat cockpit. The QRH contained a procedure for crossfeeding fuel in the case of an imbalance. This procedure contained a constraint ... a caution note that suggested the possibility of a fuel leak and specifically advised against using the crossfeed procedure in such a case. The crew did not reference the QRH to crossfeed ... who does? And why not? This is nothing but the normalization of deviance. We've all done it a thousand times, we know how to do it, and we've never had a fuel leak. It is possible that had they referenced the QRH, this constraint would have captured the decidedly unruly trajectory resulting from their decision to crossfeed into the leak.

The use of SOPs to remain within the initial space of safe operation will protect the function of procedural constraints, which function as error traps. For this reason, Figures 11.1–11.5 show the constraints as embodied in the rules, procedures, and constraints line. But this is where I have had a nagging concern with the way threat and error management is generally taught. While the idea of trapping errors is crucial, the word "trap" is almost always limited to use as a verb, an actual action. As pilots, we are supposed to trap errors. Fine; this is undeniable. As zebras, we are also supposed to run from the lion. However, the word trap also exists as a noun, describing the many constraints that Amalberti correctly located within SOPs ... zebra stripes. It must be understood and used both ways.

And right here is where the idea of envelope protection comes in. Envelope protection is nothing more than automated protection of the margins. However, much of the problem with envelope protection is that it is advertised and trained as a set of automated constraints. There is a powerful relationship between configuration, that is, airworthiness margins and the margins of aerodynamics, altitude, and performance. As the configuration degrades, for example, from normal to alternate law, the envelope protections also degrade. The automated constraints degrade. The actual margins, the ones that the envelope protections are designed to protect, *still need to be protected.* So the challenge is to understand which margins have become exposed to degraded protection and ensure they remain protected the old fashioned way. A key part of this is ensuring that the airplane as a system is properly configured, as intended by certification, to operate with degraded envelope protections. It is extremely important to understand the tight interdependency between configuration, degradation of configuration, and the margins that have become exposed.

MISTAKES WE DO NOT WANT TO MAKE

The third element of strategy development is to remain aware of lessons learned and thereby identify mistakes we do not want to make. A good number of such mistakes, which are nothing more than specific risks, are already embodied in the constraints. If we protect the constraints, that will cover most of this. However, each day is different. Each day may create conditions, an environment, in which you see a particular mistake or two, or three, that you've either made before or know about someone having made before. I don't want to get to my alternate without enough fuel to deal with yet another glitch. I don't want to get caught way out over the Atlantic with no way to get back to the coast because there's a wall of thunderstorms in the way. I don't want to overestimate the braking action.

A number of years ago, we had just resumed mainline service from Dallas to Amarillo, Texas. I had never been there. As it happened, we were scheduled to operate that leg very late at night, as the fourth leg of the day. I dutifully reviewed the approach charts and company information. Off we went. Shortly after takeoff, we were re-cleared direct to the Amarillo airport. As we got up to the top of climb, my first officer casually said, “Gee, I bet the tower is going to be closed”.

Well … yeah … I guess it will be. Of course, it will be; it was after midnight. I had missed this point, failing to connect the dots about the time of day. No big deal; we have procedures for operating into a non-tower environment. I pulled up the policy manual, dug out the non-towered procedures, and immediately discovered that for scheduled service … that was us … we were supposed to get an addendum to our flight release containing acceptable sources of weather information, presence of company personnel, etc. Which, of course, we did not have.

This was academic; we had a legal weather report and we knew the company was waiting. But we were not in compliance with the company manual. A less-than-brief discussion with dispatch on the no. 2 radio, which used up a considerable amount of my residual attention margin, led to a conclusion that all was well. That was one mistake I should not have made.

Shortly thereafter, the center controller announced that the radar coverage would not allow vectors for an approach and asked which approach we wanted. We were ready for this; we wanted the ILS 22 approach. So the controller cleared us for the ILS 22 approach. Simple as that.

But we were going direct to the airport; we had no clearance to any initial or intermediate approach fix. The mistake I did not want to make here was to just punch in a fix without a specific clearance. This took a bit of negotiating with the controller; we finally got the proper clearance.

But now there was a significant mistake that I really did not want to make. The ceiling was a solid overcast at around 1,000 feet. The approach lights were pilot-operated when the tower was closed. Decades before this, I had gotten really squeezed in a Metroliner one night, returning to Lynchburg from Charlotte. That night, despite two approaches, we could not get the approach lights turned on. Too many clicks, not enough clicks, I never was sure, but we ended up at Charlottesville with about 600 pounds of gas and the weather really coming down. Whatever had actually happened, that was a mistake I did not want to make. The first officer and I discussed this at length and made very sure we knew what we needed to do to get the bloody lights turned on.

And it worked brilliantly. We landed quite comfortably.

The next day, reviewing the approach charts, I noticed a little note on the ILS 22 chart: radar required. Hmmm … if the controller didn't think she could vector us for the approach, I'm guessing we did not have the required radar coverage. Another mistake I should not have made.

You get the idea. We caught the big ones that night, but we missed the little ones.

Figure 11.6 shows how this affects our strategy picture.

The tight curve depicting mistakes we do not want to make is a broken line, because particular specific risks will only be significant in local areas of operation, not all areas. But the idea is to build another barrier, a flight-specific constraint, that creates pressure against making a particular mistake. The upper part of the curve is much lower to indicate that some mistakes we do not want to make are associated with too great a degree of satisficing. The lower part is well above the workload limits because other mistakes that we do not want to make are related to diminishing our margin of residual attention.

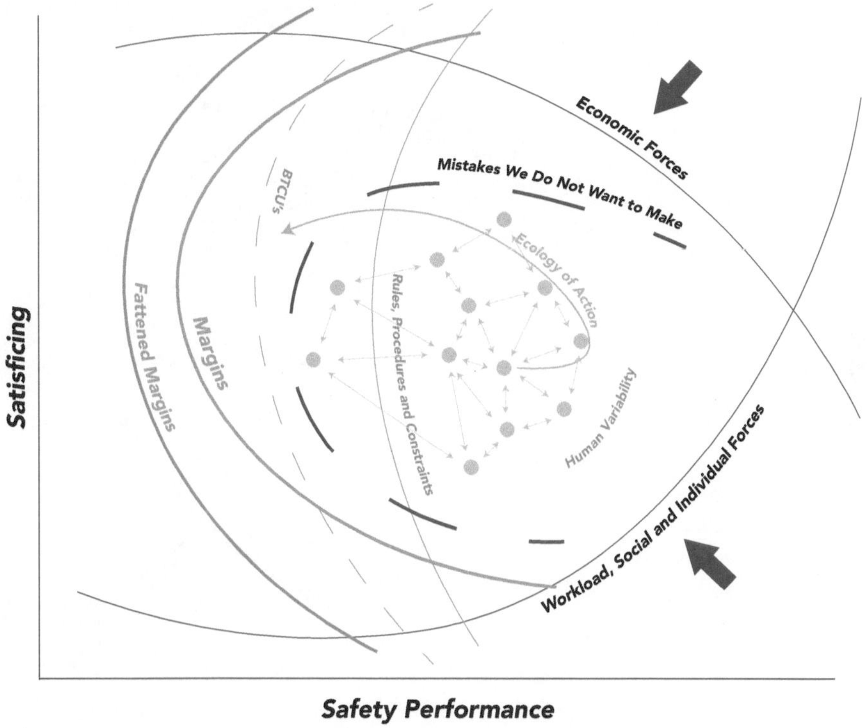

FIGURE 11.6 Space of safe operation with mistakes we do not want to make.

RELUCTANCE TO SIMPLIFY AND SENSITIVITY TO OPERATIONS

This is the fourth step in our structure of strategy. We need to add resistances reflecting a reluctance to simplify and a sensitivity to operations. We have discussed these ideas at length in Chapters 9 and 10. Figure 11.7 shows how the addition of these ideas might affect our strategy.

The curve representing these features lies outside of the curve depicting mistakes we do not want to make, because they remain probabilistic; they are not yet specific. They refer to a concept, a general part of a strategy designed to ensure that human variability provides negative feedback to drift, and does not apply positive feedback. Note that neither concept does anything to reduce workload; quite the opposite. They are again associated with reducing the degree of satisficing. The margin of residual attention has to be protected independently. However, both features impart a downward pressure, away from the wider gap between the BTCUs and rules, procedures, and constraints and toward greater efficiency.

This stage of the completed strategy will hopefully build boundaries to human variability, as well as protections against variability in the environment, that lie well inside the space of safe operation defined by the aggregate margins. With any luck … and no guarantees … we will have a strategy that optimizes our ability to manage the ecology of action.

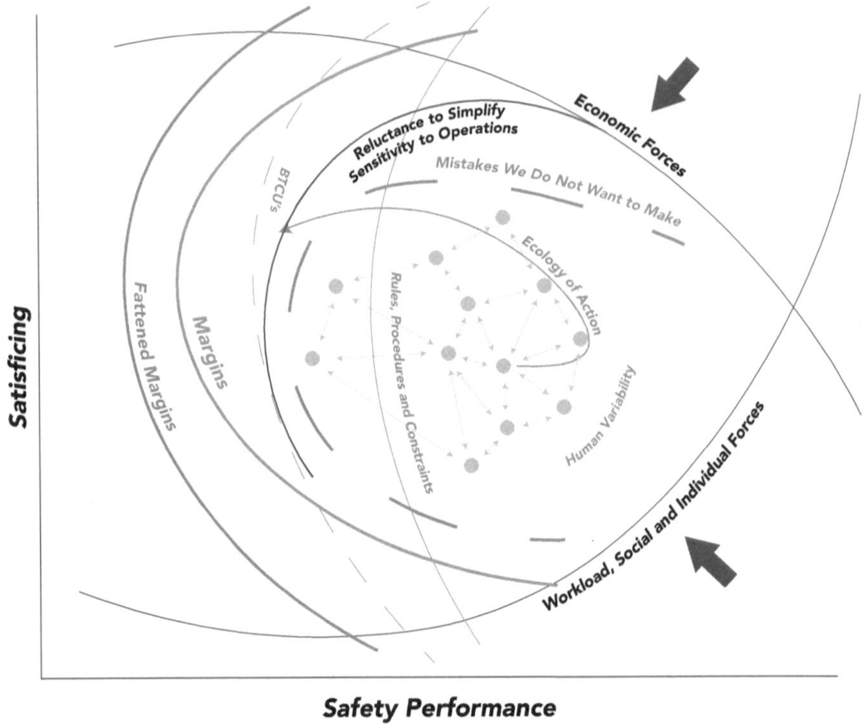

FIGURE 11.7 The space of safe operation with reluctance to simplify and sensitivity to operations.

Edgar Morin's description of an initial strategy that "elaborates a scenario of action based on an appraisal of the certainties and uncertainties, the probabilities and improbabilities of the situation"[17] is proactive and anticipatory. The appraisal is done through the first two components of prudence as defined by St. Thomas: taking counsel and judging of what you have learned. The process incorporates the principal elements of prudence … foresight, memory, understanding, reasoning, docility (willingness to learn), and caution.

Morin's subsequent description of the need to modify a strategy, "according to information gathered along the way and hazards, mishaps or good fortune encountered",[18] is also proactive, and really is nothing more than St. Thomas' idea of circumspection. Circumspection returns us to a proactive, anticipatory strategy predicated on new inputs. Shrewdness, on the other hand, the ability to engage in quick learning without the benefit of much counsel … is essentially reactive. Shrewdness opens the door to resilient performance.

RESILIENT PERFORMANCE

Resilience is another one of those words, like prudence and mindfulness, that we use all the time and, in passing, believe we understand what it means. And, like prudence and mindfulness, it's not that simple. Hollnagel has described a dichotomy in the definition of resilience; on the one hand, it is viewed as the ability to restore steady-state normalcy following a perturbation, while on the other, it is understood to mean a systemic capability to sustain normal operations during periods of rather unsteady, possibly unstable disturbances and conditions.[19]

Hollnagel has developed a framework for describing resilient performance in organizations. As with much of what we have discussed, I believe this framework can also be adapted to human performance in the cockpit, with a little massaging. Resilient performance is integrated into the modification of the strategy, but is also engaged when we are tracking an unruly trajectory and shooting it down. Hollnagel's four elements of resilient performance are Responding, Monitoring, Learning, and Anticipating.[20]

When we encounter a sudden perturbation, a disturbance, what is actually a peripeteia, a departure from the norm, we may benefit from St. Thomas' idea of shrewdness, "a habit thereby congruities are discovered rapidly".[21] We must both respond to the disturbance and monitor the result. For example, the first thing Captain Richard de Crespigny did when the engine blew up on his Airbus 380 was to punch the altitude hold button, commanding the autopilot to lower the nose. The very next thing he did was to begin monitoring, and he immediately realized that the autothrottles had not responded to the pitch reduction. So he responded again, disconnecting them and reducing power.

De Crespigny and his crew then embarked on a rather tortuous period of monitoring, responding, monitoring some more, and gradually moving into the second two elements of resilient performance: investigating, learning what had happened, learning about what that meant, anticipating how that changed the airplane's relationship with the environment, and imagining detailed next steps. In fact, this was a continuous loop. As they learned and anticipated, they continued to monitor and respond. And then learn some more.[22]

Hollnagel has explained the final element of resilient performance, anticipating, actually has two levels.[23] On the first level, the organization anticipates how its situation will evolve with respect to the outside world. On the second level, the organization begins to anticipate how its situation will change the organization internally. De Crespigny did something similar; he became aware of, monitored and managed both how the situation was changing the relationships with the environment, as well as how it was changing relationships and capabilities within the cockpit.

It is important to recognize that the first two elements of resilient performance, responding and monitoring, necessarily work together, and not necessarily in a linear sequence. They really both happen at the same time. Sadly, to a great extent, what we are taught in the simulator stops at these two elements; we respond to the engine failure with several control inputs and then we execute a checklist, while we monitor the situational awareness. Rarely is there an opportunity, or the time, to move beyond this into the aspects of learning and anticipating. These latter two elements in Hollnagel's model really comprise Karl Weick's ideas about resilience, when he describes the "capability to investigate, learn, and act *without* knowing in advance what they will be called to act upon",[24] and further specifies this as the "capacity to do quick study, to develop swift trust, to engage in just-in-time learning, to imagine detailed next steps, and to recombine fragments of potentially relevant past experience". We don't do a lot of "just-in-time" learning in the simulator, and we are rarely called upon to "recombine fragments of potentially relevant past experience". You may have noticed that this is not part of the Advanced Qualification Program.

However, resilient performance is not a module that we only pull out when an engine blows up or the pitch trim runs away. When Morin argues that the strategy must be "modified in the course of action in response to chance, information, change of context; and provides for eventual torpedoing of an action that may have taken a harmful course", he is talking about circumspection, which really uses many of the same tools as resilient performance. The change in strategy results from monitoring the situation, responding to the change, and essentially feeding back into the prudence process through investigation, learning, and imagining detailed next steps.

What is clear in this analysis is the bilateral reciprocity between strategy and resilience. While they are independent ideas, they are related through feedback. We begin constructing a strategy through the application of prudence, with foresight defining the objective. The strategy must be proactive and anticipatory, and is constructed to tolerate variability in performance and adapt to modification in the operating environment. It will be built around mistakes we do not want to make, a reluctance to simplify, and a sensitivity to operations. The margins will be strengthened as necessary. Once in operation, using ideas like mindfulness, preoccupation with failure, a reluctance to simplify, and a sensitivity to operations, we operate a continuous loop of responding, monitoring, learning, and anticipating. The third element of resilient performance, learning, engages the modification of the strategy. The fourth element, anticipating, re-activates the cycle of prudence, the final component of which is always executing command.

Anticipation involves acting before there is something to respond to. This is what prudence is all about. But as Hollnagel points out, "Anticipation makes it necessary to take a risk. Acting before there is something to respond to introduces the risk that the prediction/anticipation may have been wrong and that the effort therefore was wasted".[25] In other words, it may be necessary to take the risk of appearing overly conservative.

In Figure 11.8, we can summarize these ideas. Circumspection leads to a revised strategy through another cycle of prudence. We always, always monitor after executing command, and Hollnagel's ideas of responding, learning, and anticipating fit into the structure of prudence quite well.

EXAMPLES

Let's look at a couple of examples showing how this works.

In the last thirty years, the concept of a stabilized approach has been codified in airline SOPs. This concept is a nifty little model of how a strategy works. What is the foresight? What do we want to do with airplane? What do we have to do to make sure that this landing goes right? Land in the touchdown zone, on speed, on centerline, and more or less aligned with the centerline. What margins do we want to protect? Aerodynamic margins ... a suitable margin above stall speed. Performance margins ... adequate stopping distance, adequate go-around performance. Airworthiness margins ... none of the performance margins are any good if

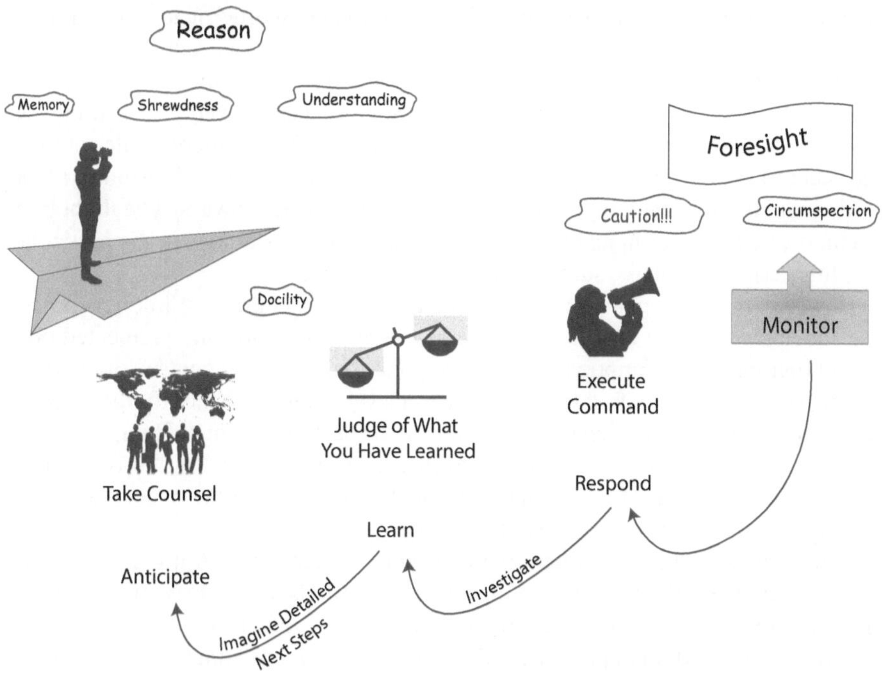

FIGURE 11.8 The reciprocity between prudence and resilient performance.

our configuration does not match the configuration that the performance data is based on. Residual attention margins … we need a stable configuration and flight path so that we have enough residual attention to spot any emergent, unruly trajectories that develop. What are the mistakes we do not want to make? Landing long, landing short, landing on the edge of the pavement, going off the side, or going off the end.

The idea is that the airplane must be configured, on speed and on glide path by 1,000 feet above the ground. It ensures that the airplane is in the best possible position to achieve the expected landing distance as well as protecting the performance margins for a go-around. By zeroing out the dynamics with a stable airspeed and sink rate, it preserves the stability of the environment and protects a margin of residual attention in the cockpit, substantially increasing the probability of error detection, while also protecting the margins of aerodynamics and performance when encountering atmospheric dynamics such as wind shifts.

The important point here is that a stabilized approach is not a program. It is not a procedure, even if it is almost invariably taught that way. It is a strategy. A stabilized approach can be achieved with or without automation, with or without a glass cockpit, and can be arrived at from a wide variety of descent profiles and lateral entries to the approach procedure. It can be achieved with or without a normal landing configuration, for example, in the case of a flap or slat failure. Nor does it necessarily lead to a smooth landing! Rather, it represents an over-all strategy to achieve a landing within the touchdown zone, on centerline and aligned with the runway, under conditions of some uncertainty, such as wind, braking action, pilot technique, even nominal fatigue. It will not completely eliminate landing incidents and accidents. But it will significantly reduce the probability of such an event.

The flight plan/dispatch release represents a strategy. It is built around foresight and follows the process of prudence. It will be constructed around such things as optimum fuel burn and optimum time enroute, but it will also incorporate routings and fuel reserves in anticipation of uncertainties such as ceiling and visibility at the destination, turbulence, icing, or even unexpectedly closed runways. The flight plan is a tailored strategy designed to deal with the known uncertainties of the particular day. It is designed to make sure that the flight goes right.

However, the flight plan is initially generated with a program. While it is proactive and anticipatory, it is built with time-sensitive information and predicated on a set of programmed assumptions. The entire point of operational control, of having a dispatcher, is to install a first level of review, a level intended to modify the strategy based on new information, change of context, or hazards not accounted for. However, the dispatcher has fifteen other flights to plan and monitor. He or she is juggling fifteen strategies, each of which needs particular attention. In my experience, they do a remarkable job. But they are not a replacement for you.

A few years back, I was operating a scheduled flight from Antigua to Miami. The van ride from the hotel to the airport in Antigua is about forty minutes long, and I usually used this time to do my daily preflight review of weather, NOTAMs, field conditions, and flight plan. This afforded me forty more minutes of time sitting under a shade tree on the beach. The roads in Antigua are not the smoothest in the world, so this can be challenging with a touch-screen device like an iPad. On this particular day, it appeared that the original strategy encoded in the flight plan might need some modification.

The convective available potential energy over the western Bahamas was substantial, over 3,300 J/kg, the precipitable water at Nassau reported by the 12:00 UTC sounding was over 1.9 inches, and there were already some cold cloud tops between Andros Island and Bimini, as well as over Nassau. To the east, the SAL had covered the Leeward Islands and eastern portion of the Greater Antilles with a few thousand feet of dust, and this layer was migrating westward. The edges of the SAL are prone to convection.

And yet, the Terminal Area Forecast that we had for Miami looked like this:

KMIA 221451Z 2215/2318 1200 P6SM SCT025 BKN250
FR230100 VRB 03KT P6SM SCT025
FM231400 13008KT P6SM SCT015

No mention of storms here. However, the Miami forecaster had this to say about the day:

> Looking at this morning's sounding, there is plenty of CAPE, before modifying the sounding. The sounding shows 2633 J/Kg. The modified sounding for a high of around 89 F, shows CAPE values nearing 5000 J/kg. The sfc to 3 km lapse rates were not too impressive once again, coming in at 4.8 C/Km. the mid level rates are much better, coming in at 6.7 C/Km. The NCAPE this morning was already at .24, and the

> modified sounding should almost .4, which is extremely high for South Florida. The 500 mb temps should be around -9C, this all adds up to any convective cells will have the ability to grow tall, and holding a decent cold pool aloft. While the low level shear is weak, at only 6, the Effective Bulk shear is 22, basically moderate. The wind aloft is not particularly strong, only around 20 kts, but should be enough to keep many cells from having their updraft killed by the downdraft. The freezing level is a bit high, at over 15k ft, so, hail is not a significant concern today.[26]

So there were two problems: circumnavigating the storms likely over the Bahamas and ensuring that we had comfortable holding and diversion fuel when we got to Miami. I had seen both problems before.

The dispatcher had already modified the program-generated flight plan, adding 800 pounds of fuel. But the standard alternates of Palm Beach and Fort Myers posed threats of their own. To get to Fort Myers, we would have to circumnavigate whatever was floating around South Florida. If we held off Miami until we were committed to Palm Beach, we could end up racing a storm up the east coast … the two airports are not that far apart and both are typically immersed in the same weather dynamic.

Time to fatten the margins. I added another 1,500 pounds of fuel.

In the end, we didn't have any trouble skirting the weather over the Bahamas. A couple of twenty-degree heading changes to the south was all we needed, and we did not need any coordination with Havana Control. When we got to Miami, my relatively new first officer was suitably impressed with the huge cell parked about fifteen miles to the west of the airport. We didn't have any problem getting in, but he could see the point; a slight variation in yin or yang, and we would have been diverting.

So what was the strategy? What did we have to do to make sure this day went right? What is the foresight looking at? Two things right up front: first, I wanted to make sure that if and when we got to the most distant alternate, I have 5,000 pounds of fuel remaining. Second, I wanted to make sure that on top of that, we had enough fuel for a fairly long circumnavigation of the possible weather over the Bahamas. It was not an either/or proposition; I wanted to be able do the long circumnavigation *and* arrive at the most distant alternate with 5,000 pounds. The uncertainty that was front and center was how the storms over the Bahamas would develop and how any storms over Florida would develop. Note that the forecaster had concluded that any cells around Miami would have the ability to develop vertically, and that the available shear would be enough to tilt the storm and keep the downdraft from killing the updraft. That is a recipe for a supercell.

In this case, the dispatcher had already added fuel to handle the enroute diversion around weather. But this was still predicated on the standardized fuel reserve, based on weight, which was usually around 3,500 pounds for the 737. I never accepted this as adequate; a simple go-around and return to land because a gear indicator light has failed will cost around 1,500 pounds of fuel. Anything below 2,000 pounds of fuel is, according to the QRH (and common sense), a fuel emergency. The margin I wanted to protect was to arrive at the most distant alternate with enough fuel, ergo time, to manage a minor malfunction during the approach without jacking up anyone's blood pressure. In other words, 5,000 pounds of fuel over the marker at the alternate for the 737. For the MD-80, I used 6,000 pounds; for the 757, 7,000 pounds.

But underlying this aspect of the strategy, the next thing we need to do to make sure this day goes right, is to protect the margins of residual attention, and in particular, the stability that Edgar Morin has pointed out will protect the function of programs. Stability in the environment is immediately threatened by any diversion. You are probably diverting because you are at or close to bingo fuel, so you don't have a lot of margin in fuel or time for additional things turning pear-shaped. Furthermore, the operating environment during a diversion is likely to be very, very busy. You won't be the only airplane diverting, and ATC will be saturated. So will the dispatcher.

The last time I had broken out of a hold east of Miami and bolted up the coast for Palm Beach, we were racing a cell up interstate 95. The plan was to come abeam of Palm Beach International Airport, make a left turn, and land straight in on runway 28R. Enroute, ATC issued a convoluted routing involving a Victor airway anchored on Freeport. This proved challenging to load into the FMC. When we got there, Palm Beach Approach Control had two other airplanes diverting down from the north and had set them up on the ILS to runway 14. We had to fly north and turn around to get in line. We had the gas, but not much more, and by the time we got on the approach, the storm had arrived, and we ended up flying through a pretty intense shower on final.

So, a diversion is wide open to the ecology of action. Once you begin the diversion, unruly trajectories, such as being vectored way out for an approach that takes much longer than you anticipated, become a serious threat. You may be diverting to an airport that you have little or no experience with, and that may not be a regular airport in your operations specifications. Once you are committed, you have few escape options. There are numerous programs that normally need to be executed during a diversion, such as changing the destination and communicating with dispatch. These are program sequences that are rarely practiced. Things like loading an obscure Victor airway route can bring about some destabilization, leading to other short-term programs getting stuck.

In the case of the Antigua-to-Miami flight, what I did not want to do is come around the end of a line of storms that pushed us halfway to Havana and realize that we did not have enough fuel to arrive at the most distant alternate with 5,000 pounds. The weather over the Bahamas had the potential to become very dynamic; I did not want those dynamics to cause any of the routine procedures to become stuck. I wanted to ensure that we had protected the constraints in those procedures and that we had the residual attention to do that. I was anticipating a lot of dynamics, and I wanted to ensure that however dynamic the flight became, within reason it remained a non-event. So, we added more fuel.

Yet there is considerable discussion possible about the cost of tankering fuel, the cost of procuring extra fuel on an island, and the statistical probability of diversion. These issues need to be respected; the company is in business to make money, and manipulating those issues directly affects that goal. That said, at the company level, all three of those issues are interpreted through a program. The company has no way of determining when that program gets stuck, nor is it their responsibility to do so. It is all yours. As Erik Hollnagel said, when you are executing the anticipatory element of resilient performance, you may have to take the risk that it will appear unnecessary.

This brings us to a critical consideration in the protection of margins. When is it acceptable to shrink the margins? Two common cases come to mind.

Suppose you arrive at your destination as planned, on time, with the pre-planned reserve fuel, etc. The dispatch release ensured that you will arrive at a weight which will allow a landing to a full stop on a dry runway within 60% of the available pavement for the most favorable runway, considering wind and weather, that was identified at the time of dispatch. This is equivalent to a 67% margin. That runway is 10,000 feet long, so you will weigh no more than the weight which will allow a full stop within 6,000 feet. But when you pull up the ATIS, they are advertising a different, shorter runway. What margins apply and what are you comfortable with?

Current operations specifications for FAR 121 require that you perform a landing distance assessment, typically if the runway is shorter than some number such as 7,000 feet. In any case, you must have at least a 15% margin. At your arrival weight, which still allows a full stop in 6,000 feet, you now only need a runway that is 6,900 feet long. Suppose the new runway you are assigned is 8,000 feet long. At your weight, that gives you a 33% margin. It's not 67%, but it's also not 15%. How much more than 15% do you think you need, based on an assessment of the local conditions?

Let's suppose you are holding, waiting for the weather to improve at your destination. You have fuel for two alternates plus reserve. It is becoming apparent that the hold is going to persist for longer than you can accept while preserving the fuel to divert to the most distant alternate. Do you delete that alternate (with dispatch concurrence)? If you do, you are cutting the margin and removing the redundancy that was originally intended to provide space for a wide range of uncertainty, such as the weather at the closer alternate.

But now that uncertainty has been diminished. From your holding pattern, you can literally see the shorter alternate and it is wide open. Does that justify shrinking the margins?

These are choices you must make, and in nearly all cases, they do not increase the risk because you have better, more current information about the uncertainties. But you must always evaluate these decisions in terms of the margins as compared to the risk. The company is going to want you to drop the distant alternate, but they are using a program to make that decision, typically based on weather and standard diversion routings. You and the dispatcher must evaluate it within the context of the strategy. What is the risk to the rest of the day going right … including all necessary procedures?

Shortly before I retired, I was operating a flight from Dallas to Chicago. There was a wicked front sweeping across the plains, and it was kicking out a rather noticeable line of thunderstorms. The dispatcher had routed us well out to the east, flying an arc out almost to Cleveland before turning back to O'Hare. He had filed us with two alternates. He also had planned us to fly at FL260 because of turbulence above and had added considerable dispatcher-add fuel. After some study of the situation, I added another 1,000 pounds.

Sure enough, as we started north around Louisville, we got another re-route. This one was way up north over the Upper Peninsula of Michigan, adding almost two hundred miles to the route.

Well, that modified the environment. Time to revise the strategy. What did we need to do to make sure this whole thing went right? The first thing we did was climb all the way up to FL 370; bumps or not, we needed to cut the fuel flows way back. Next, based on analysis en route of the weather situation, I suggested to the dispatcher that we drop the alternates we had filed and add Milwaukee, which would not see the line of weather until well after Chicago had. This required much less diversion fuel. He agreed.

We came down across the lake, looking at a very dark line of weather just a couple of miles west of O'Hare and landed with a comfortable margin of fuel that would have allowed a very sedate missed approach and diversion up to Milwaukee, which was indeed CAVU as we passed by it. It was charmingly dynamic, but it was a non-event.

The proper use of prudence … taking counsel, judging, and executing command … applies foresight to identify margins that might be a bit weak, strengthen them, identify mistakes we do not want to make, then build a cognitive map of the safe operating space. The proper use of SOPs protects margins and protects constraints designed to capture error … unless the environment becomes modified or unstable and the program becomes stuck. The possibility of stuck programs can be identified through mindfulness, preoccupation with failures, a reluctance to simplify the situation, and a sensitivity to operations. At this point, we can engage Erik Hollnagel's resilient performance loop, respond to the situation, update and monitor our situational awarenesses, learn what has changed, and anticipate how our relationship with the environment and the airplane has changed. A suitable margin of residual attention, either through slowing down, downloading the cockpit workload, and/or arresting and stabilizing the situation, that is, stopping the descent, going around, taking one more turn in the hold, will provide the time and space to identify unruly trajectories and shoot them down. When the program becomes stuck, we need to abandon it and focus on the strategy and then modify the strategy to protect critical margins and restore stability.

CREW COMMUNICATION

And in a multi-crew airplane, none of this works without effective crew communication. This is the final element of strategy. We have all, by now, had plenty of exposure to ideas like active listening and assertive statements. These are tactical necessities. We need a strategy to ensure that they work and that communication is effective and established. That begins with the crew briefing.

A crew briefing is aimed at paving the way for resilient performance, particularly a capacity to develop swift trust, just-in-time learning, and to collectively imagine detailed next steps and apply relevant past experience. It utilizes several small strategies. I always tried to accomplish a thorough crew briefing; because this starts with the cabin crew, it often entailed a literal foot race through the terminal to get to the gate before the agents started boarding. The broad strategy here is to establish a face-to-face relationship with the cabin crew and to synchronize our thinking. If I had time, a few minutes of genial banter was always a worthwhile investment. This allowed me to become familiar with personalities and individual states of mind. For example, if one of your flight attendants is brand new, or was late to sign in, or hasn't eaten all day, you know there will be some stress and distraction. In all cases, it is essential to remember that the next time you talk to that flight attendant, it will be on the interphone, the audio signal will probably be lousy, there will be no body language, shit will be hitting the fan, and both of you absolutely have to be able to interpret the communication effectively.

However, because there is often very little time available before passenger boarding, I consolidated the briefing into three main points, each with a specific purpose. The first thing I did was to remind the cabin crew that whenever we open the cockpit door, we needed two flight attendants up front, even if just handing in a meal. The strategy here was to simply remind them of the criticality of security procedures and elevate those procedures in their state of awareness; it was also to remind them of the importance of *all* SOPs. Next, I told them that I did not want to leave the gate with anyone onboard that they, the flight attendants, would rather not have onboard. The strategy here was to clearly affirm that I trusted their judgment and experience, and I was not predisposed to overrule that judgment. The simple truth is that any flight attendant over the age of twenty-one has more experience spotting problem passengers than I ever will. The third thing I said was that I kept the interphone on all the time. I reminded them to use the correct procedures for calling the cockpit and to generally respect the sterile period; but I also pointed out that if, at *any* time, regardless of the sterile cockpit rule, they truly believed that I needed to know what's going on in the cabin, all they needed to do was pick up the phone and start talking. I also mentioned that because of this, I would hear any conversation they had regarding safety in the cabin … medical issues, smoke, whatever. The strategy here was to establish us as a team, to remove a barrier to effective communication, and to reduce any anxiety they may have had about inappropriately disturbing the cockpit. Ultimately, I would make the decision about what I needed to hear and didn't need to hear. If I didn't want to listen to them, I could turn them off. But if we were on final approach and the cabin started to fill up with smoke, I did not want them to hesitate, and I did not ever want them getting confused about which buttons to push because they are qualified on seven different airplanes.

This preparation continued with a briefing with the first officer. In this case, there is almost always time for informal banter and this is crucial to establishing dialog. We would discuss the rejected takeoff procedure, I would very specifically emphasize the use of the Quick Reference Card for an evacuation, that we would be in no hurry, and I would remind him or her of their duties during an evacuation (many have not thought about this since initial training). This is, again, a reference to, and an emphasis on, the importance of SOPs.

I also established a basic framework for abnormal situations. The pilot flying would fly the airplane *and* handle radio communications. The pilot monitoring would handle the QRH … without distraction. If the situation became complicated and we felt it appropriate to change roles, particularly if it was appropriate for me, as the captain, to assume to overall supervision of the situation by giving the airplane to the first officer, we would do that. But from the outset, the pilot flying would affirm positive control ("My airplane"!) and take over the radios, while the pilot monitoring dug out the QRH and started the correct procedure.

I would also mention to the FO that at my age, I had a tendency to walk into a room, turn around, and leave without doing what I came into the room to do … I still do. This is a fact, but it is also a strategy; I wanted to set him or her up to be at ease reminding me of things that it appears I have overlooked. I wanted the first officer to be alert for my failures and fallibility; actually, I wanted him or her to *expect* such failures. More than one have saved me from serious embarrassment.

One fine day, we had departed Dallas for Colorado Springs in an MD-80. During the climb, because I had the interphone on, I heard the no. 4 flight attendant call the no. 1 flight attendant. She asked, "Do you smell that?" The no. 1 said, "Yes, I do …"

The no. 4 flight attendant was seated at the back of the airplane. The no. 1 was at the front of the airplane. Whatever they both smelled, it was all through the cabin. This was not good.

I punched into the interphone and asked, "What are you smelling?" They said a burning odor … they couldn't quite place it, maybe electrical, but they weren't sure. But, they said, they could also see a faint haze in the cabin.

The environment had immediately been modified and become unstable. Time to modify the strategy. What do we have to do to make sure this goes right? What was the mistake that I did not want to make? From the standpoint of resilient performance, we had already begun responding, updating our situational awareness, and learning something about what had happened and what that meant. What should we anticipate?

On June 2, 1983, Air Canada 797, a DC-9-32, experienced a fire in an aft lavatory. Because the captain did not establish the true extent of the fire, he delayed initiating an emergency descent. The aircraft landed at Cincinnati seventeen minutes after the fire was discovered. Within ninety seconds after the first doors were opened for evacuation on the runway, a flash fire occurred. Twenty-three passengers died. The NTSB concluded that the crew could have landed three to five minutes earlier had they promptly initiated the emergency descent, which, had they asked for more comprehensive information from the cabin crew, they would have.

That is the mistake I did not want to make.

The first officer was hand flying the climb. What margins need to be protected? Three: residual attention, airworthiness, and time to evacuate. The first thing I did was say to the first officer: "You have the airplane, you have the radios. I have the QRH. Autopilot ON, declare an emergency, start a turn back toward Dallas, and tell them we need to come straight back to the airport … right now".

Residual attention was protected by downloading, getting the autopilot on, and by assigning tasks. Next, we needed to protect the airworthiness margins. The QRH procedure was designed to do this; by executing it, we eliminated galley power and power to several fans, which are the source of ignition in the vast majority of inflight fires. Once that procedure was completed, I briefed the cabin crew and then made a PA announcement to the passengers. At no point did any programs become stuck. We had reconstructed a safe space of operation, as safe as it could be, by protecting the margins of residual attention, airworthiness, and, by getting the airplane pointed back to the airport, starting down, and establishing an emergency status with ATC, we had also protected the critical margin of time available to evacuate, if that needed to happen.

During our return, one of the flight attendants called and put a deadheading pilot on the interphone. He said that the odor and haze had dissipated. This was useful information. However, he then suggested that he thought we could continue on to Colorado Springs. I almost laughed; I said that since I'd already shut down everything in the airplane that might be a source of combustion, we had no idea why the odor had dissipated, so, no, we were not continuing the flight.

As it turned out, the haze and odor resulted from oil being ingested by an air conditioning pack. Not quite the catastrophe encountered by the Air Canada crew, but as they so sadly taught us, you don't know that, and you cannot ever compromise the margins.

NOTES

1. Edgar Morin, *Seven Complex Lessons in Education for the Future*, United Nations Cultural Organization, November 1999, 47.
2. Winston Churchill, "Sinews of Peace", Westminster College, Fulton, Missouri, March 5, 1946, International Churchill Society, https://winstonchurchill.org/resources/speeches/1946-1963-elder-statesman/the-sinews-of-peace/.
3. U.S. Civil Aeronautics Authority, *Civil Aeronautics Bulletin No. 5, Flight Instructor's Manual*, June 1, 1939.
4. Federal Aviation Administration, *Pilot's Handbook of Aeronautical Knowledge*, 2016.
5. Hans Hoermann, "FOR-DEC - A Prescriptive Model for Aeronautical Decision Making", Human Factors in Aviation Operations, Proceedings of the 21st Conference of the European Association for Aviation Psychology, Vol. 3.
6. Jens Rasmussen, "Risk Management in a Dynamic Society, a Modelling Problem" *Safety Science*, Vol. 27, No. 2/3, 1997, Elsevier Science, Ltd., 183–213.
7. Rene Amalberti et al., "Violations and Migrations in Health Care: A Framework for Understanding and Management", *Quality and Safety in Health Care*, December 2006, doi: 10.1136/qshc.2005.015982.
8. Erik Hollnagel, Robert L. Wears, Jeffrey Braithwaite, "From Safety-I to Safety-II: A White Paper", published by the authors, 2015.
9. Ibid.
10. Morin, 47.
11. Erik Hollnagel, "The Natural Unnaturalness of Rationality", Proceedings of the Seventh International NDM Conference, Editor J. M. C Schraagen, Amsterdam, The Netherlands, June 2005.
12. Nicholas Nassim Taleb, *The Black Swan,* Random House, 2007, Chapter 8.
13. United States Code of Federal Regulation, Title 14, Subchapter G, Part 121.
14. National Transportation Safety Board, Accident Number BFO88FA010, N23AZ, Swearingen SA-226TC, Chantilly, Virginia.
15. National Transportation Safety Board, File No. 1-0016, Aircraft Accident Report, Eastern Air Lines, Inc., Lockheed L-1011, N310EA, Miami, Florida, December 29, 1972
16. Rene Amalberti et al.
17. Morin, 47.
18. Ibid.
19. Erik Hollnagel, "Resilience Engineering", https://erikhollnagel.com/ideas/resilience-engineering-2004, accessed March 17, 2022.
20. Erik Hollnagel, "RAG – Resilience Analysis Grid", 2015, https://erikhollnagel.com/onewebmedia/RAG%20Outline%20V2.pdf, accessed March 17, 2022

21. Thomas Aquinas, *Summa Theologica*, translated by Fathers of the English Dominican Province, e-artnow Kindle Edition, 2013, IIa-IIae, q.49, a.4, ad.obj.3.
22. Richard de Crespigny, *QF32*, Pan Macmillan Australia. Kindle Edition.
23. Erik Hollnagel, "Resilience Engineering".
24. Karl Weick, Kathleen Sutcliffe, *Managing the Unexpected: Assuring High Performance in an Age of Complexity*, John Wiley and Sons, Kindle Edition, 2015, 94.
25. Hollnagel, "Resilience Engineering".
26. National Weather Service, Miami, FL, Forecast Discussion, June 22, 2020.

12 Case Studies

AMERICAN 1400 AND SOUTHWEST 1380

On September 28, 2007, American 1400, a McDonnell Douglas (Boeing) MD-80, took off from St. Louis with a start valve jammed in the open position on the left engine.[1] The cockpit-operated start valve had been deferred, requiring a manual operation of the start valve externally by maintenance. The valve was jammed open because the maintenance folks had departed from a standard operating procedure and used a locally fabricated tool to operate the valve. Predictably, this resulted in the starter motor disintegrating and a left engine fire immediately after takeoff. The first officer declared an emergency and began the engine fire procedure, but was interrupted by a discussion with ATC about runway availability, followed by a discussion with the captain about which runway to return to land on. Without making a decision on the runway, the captain inexplicably jumped ahead to briefing the flight attendants. The first officer dutifully offered to fly the airplane while the captain briefed the cabin crew. Once this was done, they returned to the QRH procedure. By the time the first officer got to step three of the fire procedure … moving the engine start lever to off and shutting off the fuel flow … over three minutes had elapsed since the initial fire indication. When the engine was finally shut down, the AC crosstie relay failed to open, the captain lost instruments, the thrust reverser unlocked light illuminated, the cockpit door swung open, and the fire damage led to a hydraulic failure, which led to a failure of those nose gear to extend, which led to a single-engine go-around, emergency gear extension, and finally a safe landing.

On April 17, 2018, a fan blade in the left engine of Southwest 1380, a Boeing 737, failed under load due to an undetected crack while the airplane was climbing through flight level 320.[2] The disintegration of the fan led to a structural failure of the left cowl, created significant damage to the engine nacelle, and a piece of the cowl struck a cabin window causing the window to depart the airframe. Within six seconds, the airplane was in a forty-one-degree left bank and the cabin had depressurized. The environment had instantly become highly modified, and the safe operating space had literally departed the fix.

Several things happened. In the six seconds immediately after the peak roll angle, the first officer used control inputs in all three axes to restore the airplane's attitude and closed the throttles to begin an emergency descent. The flight crew donned their oxygen masks. The captain shut off the left engine fuel cutoff switch. They declared an emergency and started a diversion to Philadelphia. The captain made a PA announcement to the passengers. Shortly thereafter, the captain assumed control as the pilot flying. The first officer began execution of the Engine Severe Damage/Fire procedure … approximately six minutes after the engine had failed. They ultimately landed at Philadelphia seventeen minutes after the engine failure, with the captain electing to use a non-standard flap setting of 5 due to the structural damage.

DOI: 10.1201/9781003369677-12

These two accidents have a lot in common and yet differ significantly. Both resulted in safe landings; with the exception of the tragic death of a woman seated next to the shattered window on the Southwest 737, there were no injuries. Both accidents were initiated by a defective mechanical component in an engine. Both actually involved some pretty respectable hand flying skills.

And in both cases, the actions performed by the crew probably did not go as well as the Safety One people would like. The Board was quite critical of the American 1400 crew, stating in the Findings that

> 13. The pilots failed to properly allocate tasks, including checklist execution and radio communications, and they did not effectively manage their workload; this adversely affected their ability to conduct essential cockpit tasks, such as completing appropriate checklists.
>
> 15. The pilots' interruption of the emergency Engine Fire/Damage/Separation checklist at a critical point prolonged the fire and led to additional problems, including the loss of hydraulic pressure, which caused the nose landing gear to fail to extend.[3]

And yet, in the Southwest 1380 report, the Board goes into great elaboration about how the crew also failed to comply with several checklists, but that in this case, those omissions were acceptable. The tenth finding (not numbered) in the report stated that

> Performing required checklists according to standard operating procedures is a critical part of safe flight operations. However, given the emergency situation aboard this flight, the flight crew's performance of most, but not all, of the items on the Engine Fire or Engine Severe Damage or Separation non-normal checklist and the nonperformance of the three other relevant non-normal checklists allowed the crew to appropriately balance the procedural requirement of executing checklists with the high workload associated with maintaining airplane control and accomplishing a safe and timely descent and landing.[4]

These are entirely ineffective and frankly, vacuous ways of analyzing what happened in both cases. About the only thing you're going to get out of the American 1400 report, as far as cockpit crew operation goes, is that you really do need to shut off the fuel in a timely fashion. Aside from that, the Board perseverated on a relationship between non-pertinent cockpit conversation and poor cockpit discipline, perhaps because they were frustrated that "no preexisting indicators in the pilots' training or performance histories were found that could explain their poor performance during the accident flight".[5] On the other hand, in the Southwest report, they more or less defined the ultimate in optimizing violation behavior, and by the way, congratulations to the crew on seeing through the mystique of proceduralization.

Indeed, what these two reports illustrate is the Board's internal conflict between the Safety One approach and the Safety Two approach. As he evolved his analysis of violation behavior, Rene Amalberti reached a somewhat similar conclusion to that which Erik Hollnagel and the Safety Two folks had reached. He observed that

> Concrete safety results are therefore the product of apparent contradictory actions: rules and constraints that guide work on the one hand and reliance on the adaptive capacities of staff in scenarios that fall outside guidelines, rules and regulations. Staff sometimes fail to follow rules in a reckless or careless manner (i.e. for no good reason) but equally often the rules are ignored because they impede the actual work itself.[6]

This passage describes the dichotomy within the industry between Safety One and Safety Two. On the one hand, failing to follow rules in any manner falls into the Safety One camp, while the "adaptive capacities of staff in scenarios that fall outside guidelines, rules and regulations" refers to the human variability that Safety Two argues is essential for things to go right as often as they do.

In their analyses, the Board argued that the American 1400 crew failed to follow guidelines, rules, and regulations. The Southwest crew, on the other hand, used their adaptive capacities in a scenario that fell outside of the guidelines, rules, and regulations. The tension between Safety One and Safety Two is implicit in the Board's findings in the American 1400 case, in which they essentially argue that the crew was careless for no good reason. Here, we have to be careful in how we interpret Amalberti's reference to a reckless and careless manner "for no good reason", as it approaches English and Branaghan's use of words like "malevolent, indolent, and hedonic", and the Board's discussion of poor cockpit discipline follows that path. That is a misdirected analysis, one really colored in by none other than the tenth axiom … a capable and competent pilot will never allow an airplane to crack up. The question that should have been asked was, why did the crew's actions seem, to them, to be reasonable and appropriate at the time?

The answer is that the American 1400 crew had no strategy. Instead, their entire response was a cobbling together of several short-term program sequences … the QRH, the flight attendant briefing, transfer of control, negotiating a runway with ATC, filling in ATC's little checklist about passengers, and fuel remaining … in fact, it was really one complete program sequence, the one they had executed in the simulator dozens of times. They were essentially ticking the boxes on a list of everything that they had practiced in the simulator.

Largely due to time constraints in the simulator, the engine failure scenario is intermixed with the engine fire scenario, and both have always placed a very high emphasis on a quick return to land. In the case of severe engine damage, that procedure is often one and the same with the engine fire procedure. Regardless of whether the failure occurs at rotation or in the second segment, the profile is essentially the same; climb to acceleration height, accelerate, clean up, establish a further climb and begin a turn, and then engage the QRH procedure. In the meantime, you dutifully declare an emergency, and the trusty sim instructor, in his role as ATC, starts peppering you with questions and issuing headings.

The result is that we cram in several QRH challenge-and-response checklists, a cabin crew briefing, a call to dispatch, and coordination with the ARFF folks, all within the ten minutes it takes to fly a normal circuit back to a shortened ILS approach. Very often, these short-term program sequences are literally boxes ticked; you tell the instructor that you are going to brief the flight attendants and he says to consider that done. This creates a very misleading mental model of what the priorities are and how those programs become stuck, leading to some potentially catastrophic consequences.

So, what is the strategy taught in the simulator? It is difficult to tell whether the quick-circuit approach is a strategy to deal with the actual failure or a strategy to get the required maneuver done before the sim period ends. In fact, the strategies for dealing with an engine fire that will not extinguish, one that will extinguish, and engine severe damage may be quite different, based almost entirely on Morin's "awareness of wager" … assessment of risk. To wit; is the engine going to burn off the wing, or is the damage so severe that we need to cautiously evaluate the aerodynamic handling qualities before attempting an approach?

The simulator template is evident from the CVR transcript. Seven seconds after the fire light illuminated, the captain said, "alright we have to go back and land". Ten seconds later, he instructed the first officer to declare an emergency. Thirty seconds after that, he told the first officer that he would fly and the first officer should run the QRH procedure. Ten seconds after that, they both began executing the procedure, disconnecting the autothrottle, and pulling the throttle to idle. Almost simultaneously, ATC offered runway 24 for a return landing. After a brief exchange with ATC, and without making a decision about the runway, the captain said, "I gotta tell the flight attendants". Fifteen seconds later, he decided to land on runway 30R. After another brief exchange with ATC, the first officer offered to fly while the captain briefed the flight attendants. Both of them were ticking the boxes of a program. Under conditions of significant stress, that was the only response they had available to them.

But in the real world that this crew was contending with, this program effectively became "stuck" due to a highly modified, and rapidly destabilizing, environment. Pretty much everything they did escaped their intentions, entered a sphere of somewhat tightly coupled interactions, and was more or less ripped out of their hands by the environment. By the time they got back to monitoring, the possibility of moving on to a learning element of resilient performance was vastly more challenging. They now had a multitude of unexpected indications; the AC bus crosstie relay did not function, the left AC bus had disappeared, the start valve light was blinking on and off, a thrust reverser unlocked light had illuminated, and the cockpit door had swung open. This was a veritable cacophony of Morin's "chance, information, change of context", and it made moving to the learning and anticipatory elements of resilient performance, then feeding back into a loop of prudence leading to a modified strategy, very, very challenging. At one point, the captain simply uttered, "This thing's fallin' apart …".[7]

The absence of strategy is further indicated by the first officer's preoccupation with the cockpit door. He made several attempts to close it, at the same time, he noted difficulty discharging the second fire bottle. This is really an artifact of proceduralization, infiltrating the program contained within the QRH procedure and further miring that procedure in needless distraction. It had zero relevance to any strategy.

The whole thing is the outcome of a model of situation management that is constructed solely from programs. What does it look like if we graph it?

In Figure 12.1, the stuck program pushed the entire body of human variability well over to the left. The variability has become unbounded. Human actions are moving outside of the margins. The ecology of action grasps one or more of these human actions and literally launches them into the sphere of interactions that add up to an accident. With one or more programs persistently stuck, all of the barriers and constraints fail.

The crew of American 1400 had no comprehensive strategy, probably because strategy is not generally a part of airline training programs. Had they actually performed this way in the simulator, they would have had a pointed debriefing from the

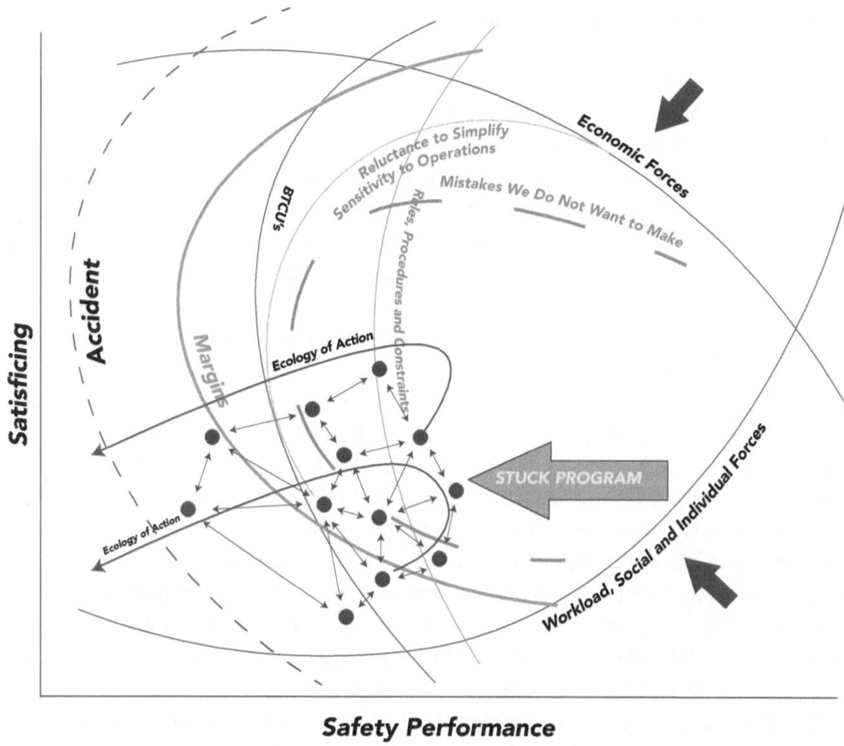

FIGURE 12.1 A stuck program.

instructor discussing the mistake of suspending the QRH procedure in order to brief the flight attendants. In fact, they probably would not have made this mistake in the simulator because the level of distraction cannot mimic the real world, and the stress experienced in the sim is nothing compared to the real deal.

Further, the simulator is not an open system. If you fail to pull the fire handle for three and a half minutes, some simulators may add some additional failures, some may not. It is highly unlikely that any simulator is going to fail the AC crosstie relay or turn on the thrust reverser unlocked light. And there is no cockpit door to unexpectedly swing open. The correction discussed in the sim debriefing may discuss the effects of a stuck program in theory, but will likely focus on reorganizing the program, ticking the boxes in the right order. The debriefing will not discuss the effects of a stuck program in any detail, because the effects of that failure within an open system cannot be modeled in a closed system. But now, in the real world, they were struggling to reassemble the program they were accustomed to in the sim, under substantially higher stress, with a bit of cognitive dissonance thrown in to boot.

What would this situation have looked like had they applied a strategy designed to protect the margins, arm the constraints, and engage human variability to remain the master of the aircraft? What margins need to be prioritized? What mistakes do we not want to make?

Let's start at the back end. The obvious mistake we do not want to make is to allow the fire to burn. The very first response to this event needed to be a clear assignment of tasks. They did that; the captain said he would fly and instructed the first officer to execute the QRH procedure. This should have done everything we want to do here; protect the margins of residual attention and protect the airworthiness margins, as well as protect constraints built into the QRH procedure. If you stop right there, you can restore stability to the environment. But the assignment of tasks was not part of a strategy to protect the margin of residual attention; it was a program, a box to be ticked, derived from simulator training. This is clear because while they vocalized it, they did not actually implement it. The first officer continued discussion with ATC, then they discussed runways, and the captain went off to brief the flight attendants.

The fire was still burning.

Churchill's idea of strategy was to provide clarity of thought. The sole focus at this point should be the protection of the aerodynamic and altitude margins, which the captain did well, and the protection of airworthiness margins as well as margins of residual attention, which continued to be compromised. Task allocation would clearly have worked here, had they implemented it. The captain can fly and talk to ATC at the same time, or he can fly and ignore ATC. You don't need permission from ATC to turn back into a downwind for a landing. In a radar environment, they can actually see you. In a non-radar environment, a simple transmission ... "We are climbing to 3,000 feet and turning right into a downwind for runway 30R. We need CFR equipment on the runway". End of transmission. The first officer needs to execute the QRH procedure *uninterrupted.* Get the engine secured, fire handle pulled, and fill the nacelle with halon. Optimize the margins for the single-engine approach ... spacing, overweight ref speeds ... and continue configuring the systems as directed by the QRH procedure. Stay focused on the strategy.

Briefing the flight attendants has no role to play here. A wise old retired Braniff captain once told me, while discussing V1 cuts in the sim briefing, "Don't worry about the flight attendants. They're still strapped in, they're not stupid and they've been trained. If they don't hear from you, they'll figure it out". And for goodness sakes, ignore the cockpit door. Don't go off on a tangent that sucks the residual attention right out of you. Protect the margins and stay focused on that strategy.

Oddly enough, it also took the crew of Southwest 1380 almost six minutes to get started on the Engine Severe Damage/Fire QRH procedure.[8] There were some important differences. First, they had lost cabin pressure, so were initially preoccupied with getting the blue side back up top ... protecting aerodynamic margins ... and getting oxygen masks on and establishing an emergency descent ... protecting airworthiness margins. They had some difficulty establishing cockpit communications due to lack of familiarity with setting up the comm panel correctly. I had the same thing happen one day when the cockpit filled up with smoke at FL 370. The first officer, who ultimately saved the day by shutting off the engine anti-ice (which turned out to be the problem), was also not familiar with setting up communications with the mask on. It's distracting; the obvious solution is to practice it frequently.

But while all this was going on, almost immediately after they began the descent, the captain shut off the fuel cutoff switch.

This is another step in protecting the airworthiness margins. The front of the engine had disintegrated. Although the captain did not know how bad the situation was yet, she did know there was a great deal of vibration and substantial yaw. She removed the fuel from the engine, depriving it of a fire potential as well as depriving it of a source of energy with which to continue tearing itself to pieces.

The industry has debated memory items, and the execution of procedure without reference to a checklist, for decades. There are pros and cons to many sides of this argument. Perhaps the most important argument against doing anything off the top of your head is that you risk losing the constraints written into the notes of the procedure … for example, once you disconnect the CSD, there's no going back, so you need the confirmation step of the procedure to act as a constraint. In this case, there was no evidence in the CVR transcript of captain obtaining the first officer's concurrence before she shut off the fuel, and there was no discussion in the report of whether the crew had confirmed this step non-verbally. Had she closed the wrong fuel cutoff switch, that could have been disastrous. That said, the difference between a program and a strategy is that the program cannot prioritize; the strategy can. There is a strong aspect of St. Thomas' idea of shrewdness here … congruities discovered rapidly. The program cannot detect an action or event entering a sphere of interactions and finally being grasped by the environment in a way that may diminish the margins. The strategy can.

Similarly to the American captain, the Southwest captain made a PA announcement to the cabin before the QRH procedure had formally been started.[9] This provided the cabin crew with assurance that the cockpit crew was conscious, made them aware of the imminent landing, and it did not materially affect protection of the margins. That said, from the comfort of my armchair, it is easy to imagine that this PA announcement could just as well have been made after they reached 10,000 feet and completed the QRH procedure. Shortly afterward, once they were below 10,000 feet, they established communication with the cabin.

About six minutes after the engine disintegrated, the captain took control of the airplane and the first officer assumed responsibility for the QRH procedure. A curious thing then happened. When the first officer asked the captain, "we're gonna need a few minutes right? To run a couple checklists? Is that right?", the captain responded, "nope just keep going", and the first officer replied, "okay".[10] This was an interesting decision. Where were they with respect to resilient performance? They had monitored, responded, and were continuing to monitor. After the captain assumed control, she had a stable airplane in the rapid descent with the engine secured. She began to enter the learning element, learning about how this situation was changing the environment and the strategy. She had begun to slow the airplane because she was concerned about the severe airframe vibration … continuing to protect the airworthiness margins. She realized that she didn't really know how airworthy the airplane was, and she was still evaluating how quickly they needed to get on the ground. She used her assessment of the airplane's behavior to judge of what she had learned … the second component of prudence … and entered the final element of resilient performance by anticipating that taking extra time to accomplish additional checklists could compromise the over-all strategy of protecting the diminished airworthiness margins.

The captain made another unique decision: she elected to land with flaps 5, rather than the procedurally recommended flaps 15 for a single engine landing.[11] This is, again, the bilateral reciprocity of resilient performance interacting with prudence to modify strategy. She was very concerned about the airworthiness margins, which she anticipated would be compromised, and in particular, based on what she had learned, she was anticipating poor handling qualities during the approach. She was concerned that flaps 15 would incur too much drag.

Both of these decisions followed the resilient performance loop through to effect a very high gain of circumspection, leading to modification of the strategy.

Subsequently, the captain updated the strategy and decided that they could afford a long final to complete the checklists. Shortly after that, the crew was apprised of the woman who had been seriously injured. At that point, the captain again modified the strategy in a way that she believed was the best chance of protecting human life. She abandoned the checklists and proceeded straight for the runway.

Figure 12.2 shows what this might look like if we graph it. At no point did the captain allow a program to become stuck. At no point did she allow a procedure to become grasped by the environment and turned on its head. Instead, her decisions strengthened and expanded the margins. She was conscious of mistakes she did not want to make. When the possibility of a stuck program became apparent, either when the first officer wanted to execute all of the checklists or later, when she had

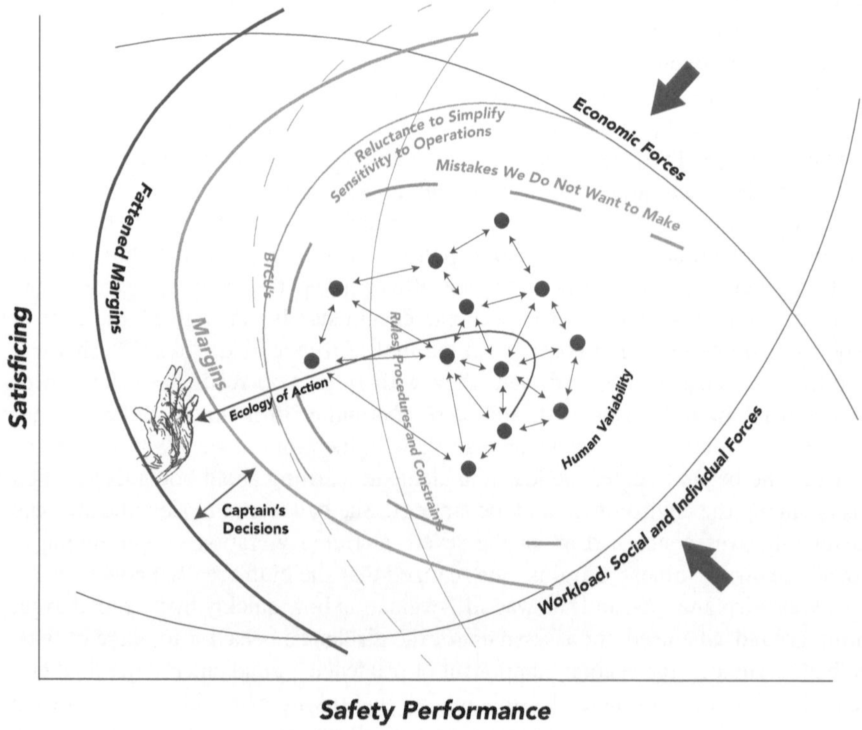

FIGURE 12.2 SW 1380 strategies and resilient performance.

opted for an extended final, she had the space within the extended margins to shoot down the unruly trajectories. Throughout the process, she was always driving toward the over-all strategy: protecting the margins and remaining the master of the aircraft.

TWA 884

Let me return to the rest of the story of our ill-fated utility bus failure enroute to Tel Aviv. There we were, climbing out of New York on a twelve-hour flight, resetting the galley busses and seeing them trip off again. What was our strategy? Again, there really was none.

A preoccupation with failure would have piqued our interest in why the utility busses dropped off when we first pushed the power up. A preservation of the constraints would have prevented us from just haphazardly resetting them on the takeoff roll. Further protection of airworthiness margins would have then driven us straight into the QRH, where we would then have attempted the single reset allowed. When that failed, further preoccupation with the failure would have indicated the need for a serious revision to the strategy, leading to appropriate circumspection.

What were the mistakes we should not have wanted to make? One would have been to not start a North Atlantic crossing with an unknown electrical fault. But the most obvious mistake no one should ever make is to continue flying *at all* with an unknown electrical fault. The system is open, the ecology of action is engaged, and we have no way of knowing what unruly trajectories may result.

Ultimately, after wringing our hands for thirty minutes, we finally tumbled to the mistake we did not want to make and decided to return to New York. But the ecology of action was still solidly engaged. We were in a hurry. In those days, with no limits on international duty time, we would have gotten back to Kennedy and either gotten the airplane fixed or gotten a new airplane … and then started out again, hours later, for our twelve-hour flight. The sooner we got back, the better.

As we approached New York, we requested an approach to runway 31L at Kennedy, as this was the longest runway and we were very much over the maximum certificated landing weight. ATC denied this request and aimed us at the ILS approach to runway 22R. We failed to adequately brief this approach and thus we failed to note that it is an offset localizer. Meanwhile, while the weather had been clear when we took off an hour earlier, a low stratus deck had rolled in off the sea and we now had a ceiling of about four hundred feet. We never noticed that on the ATIS printout. The reference speed was about 165 knots. As we commonly flew 767-200s and 757s, we failed to recall that the flaps 30 limit speed was not 162 knots, but 170 knots for the 767-300 we were flying. Thus, we got distracted by the perceived need to fly the approach at flaps 25. The upshot was that the captain was unable to align the airplane with the runway and get it on the ground in the touchdown zone; this led to a go-around from a radar altitude of zero, whereupon we put the icing on the cake by over-speeding the flaps when the controller leveled us off at 2,000 feet.

Practically every action we took escaped our intentions. We allowed time pressure to cripple any margin of residual attention. With that margin crippled, we had zero space within which to identify and shoot down any unruly trajectories. We were literally turning final on an actual instrument approach while still trying to recalculate a ref speed for a flaps 25 approach; we had no time to do that and certainly no residual attention with which to figure out that we were thinking about the wrong flap limit speed.

To put this idea of residual attention in perspective: during our missed approach and subsequent circuit around for a second try, I remembered that the ILS 22R is an offset localizer. I kept looking for that note on the approach chart. I could not find it. I did not find it *until the next day.*

Had we been truly sensitive to operations, we would have acknowledged the complexity and risk associated with an overweight landing and built decisions around an awareness of the ecology of action. This would have led to a revision of our strategy … had we had a strategy, which we did not. The easiest way to do that would have been to establish a hold for a couple of turns, protecting an adequate margin of residual attention, while we reviewed the weather, the approach, and any overweight landing considerations. A genuine sensitivity to operations would have led to us affirming a state of emergency … an overweight landing is, by default, an exercise of the captain's emergency authority … and insisting on runway 31L, thus fattening up our margins of performance. Instead, we completely failed to protect the margins of residual attention. We attempted to protect a performance margin by requesting the longest runway, but when ATC refused this, we abandoned the priority on that margin, and accepted the approach to runway 22R. We opted not to dump fuel, another step that would have protected the performance margin. With no residual attention, we missed the lately developed low ceiling and failed to adequately brief the approach. The result was a high speed, overweight go-around from a radio altitude of zero, a very unruly trajectory that we barely shot down, followed by overspeeding the flaps … another decidedly unruly trajectory.

AMERICAN 965

Let's return to American 965 at Cali. A simple strategy for approaching an airport in mountainous terrain is to never descend below the grid MORA unless you are on a published airway or approach segment … protecting the altitude margins. There is a pretty obvious mistake here that you really do not want to make. In this case, it is easy for a crew accustomed to the protections of a radar-controlled environment to interchange that environment with this one. However, this controller had no radar, no altitude readout, and he had issued the descent clearance predicated on the aircraft tracking the cleared route. He had no idea where the airplane actually was.

So what was their strategy? It really doesn't seem like they had one either. Again, they were implementing a series of short-term program sequences, such as trying to load the arrival into the FMC, and those programs got stuck when the external environment was modified. Where were they with respect to the aerodynamic margins, the altitude margins, the airworthiness margins, and margins of residual attention? What happened to their safe operating space? The change of runways almost immediately wiped out any margin of residual attention. This is abundantly clear from the cockpit dialog. Finally, in repeated attempts to engage short-term programs which persisted in becoming stuck, the environment became unstable, they lost situational awareness, and thus failed to modify the strategy to ensure protection of the altitude margins, one of the two most basic margin sets we work with.

The point of a sensitivity to operations, really the point made by St. Thomas in describing circumspection, is to be alert to Morin's unruly trajectories. It is, in fact, to be aware of Morin's ideas about deviations, transformations, chance, information, change of context, information gathered along the way and hazards, mishaps or good fortune encountered, all of the things which require a modification to the scenario of action, a change in the strategy.

Again, it is important to recall Morin's explanation of the difference between a program and a strategy:

> A program sets up a sequence of actions to be executed without variation in a stable environment, but as soon as the outside conditions are modified, the program gets stuck … We can use short term program sequences within our strategies. But for things done in an unstable, uncertain environment, strategy imposes.[12]

Short-term program sequences, such as loading a direct routing in the FMC, may be embedded within the strategy, but they are not a substitute for a strategy. Rather, they are rote procedures born from simulator demonstrations of declarative knowledge about the FMC, subsequently perfected in routine execution in order to demonstrate immediate compliance with ATC instructions. The environment that American 965 entered was clearly unstable and uncertain, and the outside conditions had been modified through multiple misunderstandings of the clearance and the FMC deletion of the Tulua VOR waypoint. The crew had quite affirmatively lost the bubble.

So what does it take to "have the bubble"? Two ideas predominate. First, there needs to be a strategy which either already protects the critical margins or is modified to do that. The modified strategy in the case of American 965 would be that we're not going to descend below the grid MORA until we are on a published segment of the routing, arrival, or approach. The modified strategy in the case of our overweight return to New York should have been to slow down the situation; slow the airplane down and slow down the process of preparation, so that a thorough consideration of the irregularities can be accomplished.

Next, go back to Rochlin's definition of having the bubble and substitute some appropriate terms: this means that the crew has been able to *construct* and *maintain the cognitive map* that allows them to integrate such diverse inputs as weather, terrain proximity, approach routing structure, the non-radar ATC environment, the controller's poor command of English (a given in Latin America), and positive positional awareness, into a single picture of the airplane's overall situation and operational status. The strategies employed are intended to facilitate the construction and maintenance of that cognitive map. Circumspection is intended to identify necessary changes to the cognitive map. A preoccupation with failure will identify possible faults in the cognitive map. A reluctance to simplify will prevent shortcuts … like going direct to a fix before even reviewing the approach, or assuming that the weather, clear an hour ago, is still clear … shortcuts that will disrupt the cognitive map, and thus lose the bubble.

As soon as that map is fragmented … as soon as you sense that you have lost the bubble … the circumspection aspect of prudence must dominate and you must utilize your command authority to stop the dynamics until the bubble is recaptured.

In the case of American 965, you must begin by stopping the descent ... regardless of the ATC clearance, until you figure out where you are. Stop the descent *now*.

UPS 1354

So, what was the strategy employed by the UPS crew? Again, as in the case of American 965, it is difficult to discern any particular strategy. Rather, also like the crew of American 965, they appeared to employ several short-term program sequences to cobble together a resolution to one of Morin's unruly trajectories. In this case, the stuck program quite clearly flew them into the ground.

The NTSB was particularly critical of the UPS crew for how they managed their rest and fatigue situation, particularly for the first officer.[13] This is all well and good, but, as is typical with the Board, they couched it in the terms of irresponsibility, a fall from grace, so-to-speak. The Board characteristically speaks from within a rather sterile laboratory, citing numerous academic studies that most crews have never heard of. Nevertheless, fatigue is *always* a factor in operations which impinge on the circadian rhythm. It is a prime example of the Safety Two argument that things that go wrong happen in pretty much the same way as things that go right. While the responsible management of rest periods is essential, it will never be perfect, and different people must manage different bodies in circumstances stressful to a greater or lesser degree. While it would be nice to get a perfect eight hours of uninterrupted sleep, and neatly align our circadian rhythms with the hour of the day, that isn't going to happen often. The prospect of fatigue impacting our performance must always be a part of strategy. The absence of a strategy for fatigue on this flight, at this time and for this approach, will leave the door wide open for fatigue to accelerate an action-based frame of mind and leave the flight crew ill-prepared to resolve the inherent cognitive dissonance.

The longer runway, with a standard ILS, was scheduled to open in twenty minutes. This was an option for the crew. However, as I pointed out in Chapter 3, pilots tend to score fairly high on the mission-oriented scale of the Blake-Mouton grid. And there was no operational reason *not* to fly the approach to runway 18. But a strategy for fatigue, modified with a sensitivity to the operation, the time of day, the recency of rest, and the like, will impose a heightened awareness of the threat posed by fatigue. The core problem is that fatigue numbs the brain to a sensitivity to operations. There is no magic technique to wish that away that we haven't already tried. But linking the strategy of a stabilized approach to a strategy for managing the immediate fatigue, primarily by fattening the margins of residual attention, can go a long way toward mitigating that threat.

Most importantly, a reluctance to simplify will lead to a more thorough, cross-checked approach briefing and setup. The Board stated that the captain "briefed the localizer runway 18 approach using the UPS Profile Briefing Guide from the UPS Aircraft Operating Manual (AOM)".[14] How they would know that is anyone's guess; more likely, he briefed it from rote memory of that briefing guide to some approximation. Obviously, that was not good enough. Approach briefings literally cry out for a sensitivity to operations; each briefing is different based on today's conditions. A sensitivity to operations means that you have to be reluctant to simplify, and a reluctance to simplify means that the rote briefing, in which you simply read off the major features of the approach chart, is not adequate. And yet, because of the rapid-fire repetition of approaches during training in the simulator, such a simplified briefing style becomes calcified. This will be the ILS to runway 13, effective November 11, 2006 … yada-yada-yada.

Again, what about a preoccupation with failure? There were two opportunities. First, when the autoflight system failed to capture the profile; this was a clear indication that something was not right with either one or both situational awarenesses. This was the situation we were faced with at Monterrey. It was after midnight, we were tired, frustrated, and practically everything we had touched had turned pear-shaped. But something was wrong, and descending without understanding what was wrong with our situational awareness was a key mistake we did not want to make.

The second opportunity came with the EGPWS warning. Part of the problem was the phrase vocalized by the system … "SINK RATE". This is not an inherently alarming vocalization. In fact, the EGPWS did not actually vocalize the phrase "TOO LOW – TERRAIN" until *after* the airplane hit the trees. Well, thanks for that. As such, this opportunity to notice failure was really fuzzy, particularly for a fatigued crew. UPS had also failed to activate the automatic callouts of five hundred feet and "approaching minimums" that the system was capable of. Either could have made a difference; they would have put the phrase "SINK RATE" in a much clearer context.

But way, way before any of that, a basic application of prudence … taking counsel … would have spotted the NOTAM regarding the closed runway before they even left the stand at Louisville. They could have modified their strategy right there.

ETHIOPIAN 302

Let's finish by revisiting the Ethiopian 737 Max accident.

What was the strategy? Again, there was none. This crew was completely trapped within the cybernetic view of automation. They had been trained within an absurdly literal version of the ultra-safe model of risk, with no allowance whatsoever for Amalberti's idea of "adaptive capacities of staff in scenarios that fall outside guidelines, rules and regulations". The program was hopelessly stuck. See Figure 12.1 above.

From the information available in the cockpit, how would you determine whether the situation was inside or outside guidelines or rules? How adaptive do you need to be in this case? And where on earth has the initial space of safe operation gone? Where are the margins that define it?

Stepping back to some of the mnemonics, what is the meaning of the words "aviate" or "fly" in this situation, five hundred feet above the ground at takeoff thrust with the stick shaker rattling and a half dozen bizarre indications flashing in your face? Could not engaging the autopilot, possibly as a means of reducing cockpit workload, be a reasonable strategic step, a step directed toward aviating?

If we are operating within a closed system, using a cybernetic approach, then at that precise moment, we have departed the ultra-safe system. The idea expressed in the ultra-safe model of risk, that there is "no need to rely on exceptional expertise even when dealing with emergencies", has completely imploded.

On the other hand, if we recognize the system to be an open system, we can recognize the space of safe operation as that well within the margins. When we invoke the idea of "fly the aircraft", what we mean is to protect the aerodynamic and altitude margins … angle of attack, bank angle, G loading. As it happened, at the moment the crew was trying to engage the autopilot, the aircraft was well within the safe space of operation, climbing at the correct attitude with ample thrust. All that needed to be done was maintain the pitch attitude and thrust, and they would have remained well within a safe operating space.

What is a key mistake that we do not want to make? Changing the configuration when you are uncertain of the airworthiness margins. Don't change the configuration until you have had time for circumspection, to become aware of the deviations and transformations, and to modify the scenario of action. Don't mess with the autopilot, and for goodness sakes, don't retract the flaps yet.

What margins do we want to protect next? We have a good rate of climb, so performance is in good shape. What about airworthiness? The stick shaker itself is a clear indication that the airworthiness of the airplane is threatened. Other indications in the cockpit are pointing at degraded capability in automation, a degraded configuration. Let's get the automation out of the picture; leave the autopilot alone and disconnect the autothrottles. Protect the aerodynamic and altitude margins manually.

With the aerodynamic, altitude, performance, and remaining airworthiness margins protected, we need to identify the correct standard operating procedure and reinforce the safe operating space with the margins and error traps that the procedure will contain. But because the primary margins have been protected and the degraded capability has been stabilized, there is no rush to figure out what's going on.

When I performed this exercise in the simulator, as part of the 737 Max requalification, this was a no-brainer. I knew what was going to happen; that was the point. We simply invoked the QRH procedure for unreliable airspeed. In the way of resilient performance, we monitored, recognized the situation, and I responded by setting the pitch at ten degrees and the power at 85% N1. There wasn't any point in going into the learning and anticipating phase, nor did I feel any need to anticipate next steps or recombine fragments of experience. We were in the simulator; we had already done all the learning and anticipation in the briefing room. That said, let's not underestimate the wild confusion, alarm, and overwhelming cognitive dissonance that can result from such an event. With all of the warnings flashing, with the agitating noise of the stick shaker, it will appear as though there are programs screaming to be implemented. There will be a powerful drive to find the security of a program, any program. But think about how strategy, constructed from an organized protection of the margins, constraints, and an awareness of mistakes we do not want to make, can prevail as the programs become stuck.

NOTES

1. National Transportation Safety Board, AAR-09/03, In-Flight Left Engine Fire American Airlines Flight 1400 McDonnell Douglas DC-9-82, N454AA St. Louis, Missouri, September 28, 2007.

2. National Transportation Safety Board, AAR-19/03, Left Engine Failure and Subsequent Depressurization Southwest Airlines Flight 1380 Boeing 737-7H4, N772SW Philadelphia, Pennsylvania April 17, 2018.
3. National Transportation Safety Board, AAR-09/03, 67.
4. National Transportation Safety Board, AAR-19/03, 85.
5. National Transportation Safety Board, AAR-09/03, 54.
6. Charles Vincent, Rene Amalberti, *Safer Healthcare, Strategies for the Real World*, Springer Open, 2016, doi: 10.1007/978-3-319-25559-0.
7. Ibid., 124.
8. National Transportation Safety Board, AAR-19/03, 3.
9. Ibid.
10. Ibid., 4.
11. Ibid., 5
12. Edgar Morin, *Seven Complex Lessons in Education for the Future*, United Nations Cultural Organization, November 1999, 47.
13. National Transportation Safety Board, AAR-14/02, Crash During a Nighttime Non-precision Instrument Approach to Landing UPS Flight 1354 Airbus A300-600, N155UP Birmingham, Alabama August 14, 2013, 88.
14. Ibid., 3.

Conclusion

The future of commercial aviation is inextricably linked to the debate between the cybernetics approach to automation and the capability approach, between Amalberti's normative approach to training, in which aviation operations can be entirely specified through standardized procedures, programs, schedules, rules, etc., and his ecological approach to training, in which aviation operations *cannot* be completely specified through standardized procedures and programs. This is nothing but a debate between a closed system and an open system, a debate over whether total knowledge of the world is achievable, over whether, in fact, we can actually think of a thought before we think of it.

Nowhere in the definitions of the cybernetic approach to automation, or the normative approach to training, or even the ultra-safe model of risk, has anyone bothered to locate responsibility.

All of these approaches to safety are really predicated on Herbert Simon's 1947 idea, that "two persons, given the same skills, the same objectives and values, the same knowledge and information, can rationally decide only upon the same course of action".[1] This is the definitive statement of determinism, wholly invested in a form of Dekker's naïve Newtonian scientism, which, in 1947, was all the rage. Simon had gone on to say that "administrative theory must be interested in the factors that will determine with what skills, values, and knowledge the organization member undertakes his work",[2] which, of course, is the entire point of proceduralization. Embedded within the description of the ultra-safe model of risk is the idea that "comprehensive preparation and training allows all operators to meet the required standards of performance and be skilled to the point that they are interchangeable within their respective roles".[3] The "interchangeable" term in that phrase is a direct reflection of Simon's original statement regarding two rational people. When the description goes on to say that the ultra-safe model "relies heavily on external oversight and the control of hazards which makes it possible to avoid situations in which frontline staff are exposed to exceptional risks",[4] it is referring directly to Simon's administrative theory.

The cybernetic approach to automation is simply a digitization of the same principles. The normative approach to training is constructed from the same foundation as Simon's administrative theory, with one notable caveat: not only will safety improvement result from more specification (through administrative theory) but also from more discipline on the part of the operators.

This was the conundrum of the Victorian era sentiments that were reflected in the tenth axiom. Becoming airborne in the early 20th century pitted human free will and decision-making against uncertainty in a more immediate and sudden manner than anything humanity had done before. As the old saw goes, "Aviation in itself is not inherently dangerous. But to an even greater degree than the sea, it is terribly unforgiving of any carelessness, incapacity or neglect". The idea of the capable and competent aviator engendered a total knowledge of the world, a perfect rationality, a human will with which to extrude certainty from uncertainty, and the words "will never allow an airplane to crack up" cemented the deal by encoding a perfect discipline.

DOI: 10.1201/9781003369677-13

The normative approach to training needs more discipline from the operators and often becomes fixated on violation behavior, because administrative theory alone can't get the job done. It turns out that two persons *can* rationally decide upon different courses of action. But far more importantly, it turns out that within the actual open system of the real world, the ideas of the cybernetic approach to automation, the normative approach to training, and the ultra-safe model of risk completely implode in the face of complexity and emergence.

When St. Exupery said that

> Nevertheless, his essential problems are set for him by the mountain, the sea, the wind. Alone before the vast tribunal of the tempestuous sky, the pilot defends his mails and debates on terms of equality with those three elemental divinities.[5]

he was describing an open system, a complex system capable of emergence. Ideas such as the cybernetic approach to automation and the ultra-safe model of risk assume a superiority to St. Exupery's elemental divinities; that is the single most important mistake that we must never make. We can, at best, debate on terms of equality; the elemental divinities, which today must include automation, are an open system, and so are we.

The answer to this debate is to deploy human variability within rationality, to locate the responsibility to use that variability with the pilot, and to vest the pilot with the authority to execute that responsibility. As Rene Amalberti said, "The crew cannot have a second role because the members continue to be legally and psychologically responsible for the flight. They must keep the final authority for the system".[6] I would add that the responsibility is not just a legal and psychological responsibility, but a moral responsibility. It is a towering moral responsibility, and it can reside nowhere else.

It turns out that responsibility is not particularly deterministic. It is not easily measured numerically nor can it really be controlled. In the systems approach to safety, responsibility remains something of a black box, more or less occupying the same spot as the idea of more discipline from the operators, embedded within the normative approach to training. It is quite possible that responsibility is a weakly emergent behavior, predictable in principal but not in specificity, another example of a combination of upward emergence and downward causation, another case of the emergence of a mental, sentient property that cannot be understood using fundamental reductionism.

But at the end of all this discussion, responsibility hasn't gone anywhere. It is still encoded in the operating rules. It is still very much a part of Cicero's idea that "True law is right reason conformable to nature, universal, unchangeable, eternal, whose commands urge us to duty, and whose prohibitions restrain us from evil".[7] With the possible exception of technical rock climbers, nowhere in any human endeavor do human beings place a greater trust in any one person than the trust placed in the crew by the passengers aboard an airplane.

The single most important question, as we hurtle toward greater automation and more sophisticated socio-technical systems, is how we execute responsibility within the structure of those systems. Like St. Exupery's tempestuous sky, these are complex systems, extremely sensitive to initial conditions. Our actions are always, always open to Morin's ecology of action. The keys to effecting responsibility lie with strategy to prevail over programs, with the protection of the margins, the preservation of the constraints, and the awareness of mistakes we do not want to make. We must practice a continuous cycle of prudence, of mindfulness, and always be reluctant to simplify and remain sensitive to operations. As it has always been, our procedural knowledge must be backed up by the most comprehensive declarative knowledge we can obtain. We must question that knowledge, test it, try to falsify it, be alert for the normalization of deviance, and never "mistake the change of a feeling of doubt into a feeling of assurance *as knowledge*".[8]

Safety is a dynamic non-event.

People ranging from St. Thomas of Aquinas to Edgar Morin to Karl Weick and Erik Hollnagel have given us the best tools that human thought can create with which to debate the elemental divinities. We are not systems managers; we are managers of uncertainty. Think again, and think every working day, of our definition of airmanship:

> Airmanship is the application of both prudence and mindfulness so as to always remain the obvious master of the aircraft, and to construct, modify and execute the necessary strategies to ensure that the safe outcome of the flight is never manifestly in doubt, while always protecting the margins in anticipation of uncertainty.

I began the journey of working this out over forty years ago, on that awful day in 1978 which took the life of my best friend. I sincerely hope that I have been effective at conveying all that I have learned and that it will serve you in good stead. Throughout your career, you will always be struggling within an environment of possible actions that are influenced by broader, enveloping social, personal, and organizational forces. It is up to you, as the pilot-in-command, to manage them.

Always remember one thing above all else: the only thing that really matters is that nobody dies.

NOTES

1. Herbert Simon, "The Proverbs of Administration", *Public Administration Review*, Vol. VI, No. 1, Winter 1946, 64.
2. Ibid., 64.
3. Charles Vincent, *Rene Amalberti, Safer Healthcare, Strategies for the Real World*, Springer Open, 2016, doi: 10.1007/978-3-319-25559-0.
4. Ibid.
5. Antoine de St. Exupery, *Wind, Sand and Stars*, Reynal and Hitchcock, 1939, 34.
6. Rene Amalberti, "Automation in Aviation: A Human Factors Perspective", in Daniel Garland, John Wise & David Hopkin, Editors, *Aviation Human Factors*, Lawrence Erlbaum Associates, 1998, 173–192.
7. Marcus Tullius Cicero. *The Collected Works of Cicero: The Complete Works PergamonMedia (Highlights of World Literature)*. Pergamon Media. Kindle Edition.
8. Karl Weick, Kathleen Sutcliffe, *Managing the Unexpected: Assuring High Performance in an Age of Complexity*, Wiley, 2015. Kindle Edition, 116.

Index

E

F

G

H

I

K

L

M